Design and Synthesis of N-Heterocyclic Donor Based TADF Emitters for Application in OLEDs

Zur Erlangung des akademischen Grades eines

DOKTORS DER NATURWISSENSCHAFTEN

(Dr. rer. nat.)

von der KIT-Fakultät für Chemie und Biowissenschaften

des Karlsruher Instituts für Technologie (KIT)

genehmigte

DISSERTATION

von

M. Sc. Zhen Zhang

aus Henan, China

Dekan: Prof. Dr. Manfred Wilhelm

Referent: Prof. Dr. Stefan Bräse

Koreferent: Prof. Dr. Uli Lemmer

Tag der mündlichen Prüfung: April 22, 2020

Band 92
Beiträge zur organischen Synthese
Hrsg.: Stefan Bräse

Prof. Dr. Stefan Bräse
Institut für Organische Chemie
Karlsruher Institut für Technologie (KIT)
Fritz-Haber-Weg 6
D-76131 Karlsruhe

Bibliographic information published by the Deutsche Nationalbibliothek

The Deutsche Nationalbibliothek lists this publication in the Deutsche Nationalbibliografie; detailed bibliographic data are available in the Internet at http://dnb.d-nb.de

ISBN 978-3-8325-5167-4
ISSN 1862-5681

Logos Verlag Berlin GmbH
Georg-Knorr-Str. 4, Geb. 10
12681 Berlin
Tel.: +49 030 42 85 10 90
Fax: +49 030 42 85 10 92
INTERNET: http://www.logos-verlag.de

Preparedness ensures success and unpreparedness spells failure.

凡是豫则立，不豫则废

Book of Rites · Doctrine of the Mean

Honesty Declaration

This work was carried out from October 1st 2016 through March 11th 2020 at the Institute of Organic Chemistry, Faculty of Chemistry and Biosciences at the Karlsruhe Institute of Technology (KIT) under the supervision of Prof. Dr. Stefan Bräse.

Die vorliegende Arbeit wurde im Zeitraum vom 1. Oktober 2016 bis 11. März 2020 am Institut für Organische Chemie (IOC) der Fakultät für Chemie und Biowissenschaften am Karlsruher Institut für Technologie (KIT) unter der Leitung von Prof. Dr. Stefan Bräse angefertigt.

Hiermit versichere ich, ZHEN ZHANG, die vorliegende Arbeit selbstständig verfasst und keine anderen als die angegebenen Hilfsmittel verwendet sowie Zitate kenntlich gemacht zu haben. Die Dissertation wurde bisher an keiner anderen Hochschule oder Universität eingereicht.

Hereby I, ZHEN ZHANG, declare that I completed the work independently, without any improper help and that all material published by others is cited properly. This thesis has not been submitted to any other university before.

Table of Contents

Abstract

As one of the most promising electroluminescent technologies, organic light-emitting diodes (OLEDs) have attracted more and more attention because of their enormous and unique potential for application in industrial products such as energy-saving lighting sources, high-resolution flexible screens and ultrafast responsive displays. Due to environmental and economic concerns, purely organic thermally activated delayed fluorescence (TADF) molecules are considered as the ideal emitters to expand the practical applications of OLEDs. Therefore, the presented thesis is focused on the development of new TADF emitters for OLEDs. Particularly, emitters for blue TADF emission, which are strongly needed for industrial applications, are thoroughly investigated *via* the evaluation of newly designed and modified N-heterocyclic donors.

In the first part, TADF emitters based on N-heterocyclic donors and boron-based acceptors were designed and synthesized. Density functional theory (DFT) computations indicated that the target molecules possess small ΔE_{ST} in the range of 0.04–0.18 eV. Photophysical studies demonstrated blue and deep blue emission with photoluminescence maxima (λ_{PL}) of 410 nm to 491 nm through the variation of the N-heterocyclic donors. A second triarylboron-based unit was also evaluated as acceptor for TADF emission. Through changing the donor units, photoluminescence quantum yield (PLQY) as high as 93% with a very short delayed lifetime of 3.1 µs was obtained.

Secondly, a series of acceptor-donor-acceptor TADF emitters based on the indolocarbazole donor and various di-phenyltriazine acceptors were evaluated. OLEDs with the EQE of over 22.1% and sky-blue emission of 483 nm was obtained. This was determined to be caused by a nearly complete horizontal orientation of the emitter. The introduction of *tert*-butyl groups to the triazine acceptor was proven to play a key role in enhancing the horizontal orientation to a high degree. The influence of steric hindrance on the phenyl bridge between donor and acceptor and electron-withdrawing strength of acceptors was further explored in this part. Additionally, it was confirmed that a dimerization strategy is an effective method to tune the performance of TADF emitters.

Lastly, electron-withdrawing groups were introduced to the [2.2]paracyclophane-derived carbazolophane donor unit (**Czp**), which possesses through-space interaction and exceptional steric hindrance to deepen the HOMO level of the emitters. This leads to a larger optical gap and a blue-shift of TADF emission. For decreasing the ΔE_{ST} to harvest a more effective TADF emission, a through-space donor dimer using the dicarbazolophane (**DCCP**) and the respective final emitters were also synthesized and used for TADF emitters, yielding blue TADF emission.

Kurzzusammenfassung

Als eine der vielversprechendsten Elektrolumineszenz-basierten Technologien haben Organische Leuchtdioden (OLEDs) stetig mehr Aufmerksamkeit durch ihr enormes und einzigartiges Potential für industrielle Anwendungen als energiesparsame Lichtquellen, hochauflösende, flexible und ultraschnell reaktive Displays auf sich gezogen. Aufgrund von Umwelt- und Kostenbedenken gelten rein organische thermisch aktivierbare verzögert fluoreszierende (engl. TADF) Moleküle als ideale Emitter um die praktische Anwendbarkeit von OLEDs weiter auszudehnen. In diesem Zusammenhang konzentriert sich diese Arbeit auf die Entwicklung neuartiger TADF Moleküle für OLEDs. Insbesondere ist die TADF Emission im (tief)blauen Spektralbereich, welche für industrielle Anwendungen von starkem Interesse. Daher wurde diese mittels Evaluation von neu konzipierten und modifizierten N-heterozyklischen Donoren untersucht.

Im ersten Teil der Arbeit wurden TADF Emitter basierend auf N-heterozyklischen Donor- und Bor-basierten Akzeptorgruppen konzipiert und synthetisiert. DFT Berechnungen deuteten darauf hin, dass die Zielmoleküle geringe ΔE_{ST} Abstände im Bereich von 0.04–0.18 eV aufweisen. Photophysikalische Untersuchungen zeigten, dass blaue und tiefblaue Emissionen mit Emissionsmaxima (λ_{PL}) zwischen 410–491 nm durch die Variation der N-heterozyklischen Donoren erreicht werden können. Ein zweiter Triarylbor-basierter Akzeptor wurde ebenfalls im Hinblick auf TADF Eigenschaften untersucht. Durch die Variation von Donoren konnten Emitter mit PLQY bis hin zu 93% und verzögerten Emissionsdauern von 3.1 µs hergestellt werden.

Im zweiten Teil der Arbeit wurde eine Reihe von Akzeptor-Donor-Akzeptor TADF Emittern basierend auf dem Indolocarbazoldonor und verschiedenen Di-phenyltriazinakzeptoren untersucht. OLEDs mit einem EQE von über 22.1% und blauer Emission von 483 nm wurden hergestellt. Als Ursache wurde die nahezu vollständige Präferenz für horizontale Orientierung des Emitters festgestellt. Das Einführen von *tert*-Butylgruppen zum Triazinakzeptor erwies sich als Hauptgrund der starken Erhöhung zur horizontalen Orientierungspräferenz. Der Einfluss der sterischen Hinderung auf die Phenylbrücke zwischen Donor und Akzeptor, als auch der elektronenziehende Charakter bei Akzeptormodifikation wurden für dieses Emitterdesign weiter untersucht. Zusätzlich wurde nachgewiesen, dass ein Dimerisierungsansatz für diesen Emittertyp eine effektive Methode ist, um die Leistungsfähigkeit der TADF-Emitter weiter beeinflussen zu können.

Schließlich wurden elektronenziehende Gruppen zur [2.2]paracyclophanbasierten Carbazolophan-Donoreinheit (**Czp**), welche räumliche elektronische Wechselwirkung und einzigartige sterische Hinderungseigenschaften besitzt, eingeführt um das HOMO Niveau abzusenken. Dies führte zu einer Erhöhung der optischen Bandlücke und damit zu einer Blauverschiebung der jeweiligen TADF-Emitter. Zur Verminderung der ΔE_{ST} Lücke, um eine noch effizientere TADF-Emission zu erreichen, wurde ein räumlich elektronisch wechselwirkendes Dimer, nämlich das Dicarbazolophan (**DCCP**), als Donor und die jeweiligen Emitterstrukturen hergestellt, welche TADF-Emission im blauen Spektralbereich zeigten.

1   Introduction

"Organic chemistry has literally placed a new Nature beside the old. And not only for the delectation and information of its devotees; the whole face and manner of society has been altered by its products."[1] was stated by R. B. Woodward to describe the power of organic chemistry. One example of this is organic light-emitting diodes (OLEDs).[2-3] In contrast to inorganic semiconductors based LEDs, OLEDs employ organic materials to generate light for general illumination and display applications.[4] Due to the absence of crystalline materials that are used in inorganic devices, OLEDs can be fabricated through vacuum deposition as well as solution-based processes like spin-coating and inkjet printing, which is helpful in achieving large-size and high throughput fabrication, and production costs savings.[5] Furthermore, organic synthesis and derivatization of molecular structures offer a delicate tool for careful tuning of molecular properties in order to modulate precisely the emitter and layer characteristics to attain the full color spectrum.[6]

In addition, OLED-based displays also possess some unique features such as high efficiency and brightness, and offer a truly black display, high contrast and saturation ratio, a low driving voltage, flexibility and transparency.[7-8] Therefore, OLEDs have attracted huge and continual attention in the organic optoelectronic materials field due to their enormous and unmet potential for application in industrial products such as energy-saving lighting sources, high-resolution flexible screens and ultrafast responsive displays for augmented reality and holographic projection (Figure 1).

Figure 1. OLED applications: OLED smartphone (left, copyright: Huawei),[9] transparent display (central, copyright: LG Display),[10] flexible display (right, copyright: cynora GmbH).[11]

1.1 Organic Light-Emitting Diodes (OLEDs)

OLEDs are a very promising new electroluminescent technology, which can effectively, controllably and economically convert electricity into light. The first example of electroluminescence was reported by H. J. Round in 1907, when he submitted silicon carbide to an electrical current of 10 V.[12] In 1955, Bernanose and co-workers reported electroluminescence for phosphorescent organic dyes, when they were doped into a polymer matrix and driven by a dielectric cell.[13] In 1963, Pope *et al.* started to investigate the basic mechanism of electroluminescence after they observed this phenomenon on anthracene crystals.[14] The initial breakthrough towards OLEDs and commercial application was achieved in 1987 by Tang and Van Slyke who fabricated the first well-designed OLED device.[2] For this, 8-Hydroxyquinoline aluminum (Alq_3) was vacuum-deposited in a layered device architecture between two electrodes. Green emission at 550 nm was generated at driving voltage of less than 10 V, with a brightness of over 1000 cd m^{-2} and an external quantum efficiency of 1%. With these ground breaking results, OLED officially steps onto the historical stage and has been given continual and major attention ever since.

1.1.1 Working Principle and Architecture of OLEDs

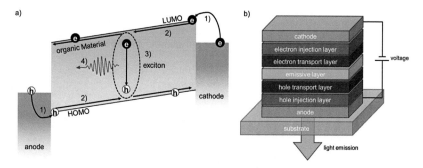

Figure 2. a) Simplified working processes of an OLED: 1) injection of charge carriers, 2) transport of charge carriers, 3) generation of excitons, 4) light emission. b) OLED architecture. Images taken from Hundemer.[15]

The simplified working processes of a single-stack OLED are shown in Figure 2.[16-17] Firstly (step 1), electrons and holes are injected into the organic materials stack from the cathode and anode, respectively. In order to realize effective injection, it is important that the energy level of the highest occupied molecular orbital (HOMO) of the organic material should be slightly lower than the work function of the anode, while the energy level of lowest unoccupied molecular orbital (LUMO) needs to be slightly higher than the work function of the cathode. Secondly (step 2), the injected electrons and holes migrate to their corresponding electrodes directed by the external electric field. In the third step (3), when the distance between the migrating positive and negative charges is closer than the Coulomb radius, recombination between an electron and hole occurs and excitons are generated. Lastly, excitons – also called molecules in excited states – tend to release the excess of energy to transfer back to their energetic ground state through radiative and nonradiative decays. When the relaxation occurs *via* the radiative decay photons emit, naming this process electroluminescence (step 4).[18]

Since the transport rates of electrons and holes are different in organic materials, normally the transport rates of holes are known to be faster than for electrons, a multilayered architecture with optimized functional materials and thicknesses for the electrons and holes, respectively, is found to be optimal in order to obtain balanced transport and recombination rates for highly efficient OLED devices.[19] An exemplary multilayer OLED architecture is described in Figure 2. In addition on a device level, an efficient out-coupling of the light is another critical factor to increase the useable external quantum efficiency (EQE) of OLED devices.

Substrate

The aim of using substrate in OLED device is to carry and support the whole OLED stack architecture. Generally, highly transparent glass or polymers are employed for smooth out-coupling of generated light. According to the different requirements and applications, the substrate is commonly rigid, but can also be a flexible film.[20]

Anode and Hole Injection Layer (HIL)

In general, high transparency is also needed for the anode in order to allow the out-coupling of light because the opposite electrode, namely the cathode, is often served by base metals or alloys which are opaque. For OLED devices, indium tin oxide (ITO, $(In_2O_3)_{0.9}(SnO_2)_{0.1}$) has been proven to be an ideal candidate as the anode material because of its appropriate energy level and high transparency.[21] Since the interface of the anode layer is typically rough, a hole injection layer

(HIL) is often coated to make the interface flat and smooth, which can also facilitate the hole injection from the anode to the following layer by aligning their HOMO levels. A mixture of poly(3,4-ethylenedioxythiophene) and polystyrene sulfonate (PEDOT:PSS) is often used in this layer to satisfy these requirements.[22]

Hole Transport Layer (HTL) / Electron Blocking Layer (EBL)

The HTL is introduced to facilitate the transport of positive charges, namely holes, to the final recombination zone in the emissive layer (EML).[23] Since the recombination of holes and electrons should be strictly limited in the EML to obtain high efficiency, this layer also serves as an electron blocking layer (EBL) for the electrons migrating from the cathode.[24] Electron-rich aromatic molecules such as carbazole or triarylamine derivatives, are often used to construct HTL materials.

Emissive Layer (EML)

This layer is crucial to the whole OLED device. The migrated holes from the anode and electrons from the cathode recombine in this layer to generate excitons. In order to facilitate effective recombination, the holes and electrons should match each other in quantity as an excess of one or the other will introduce reduction, or oxidation-based degradation of the materials.[25] This layer commonly consists of a host/guest (emitter) system to inhibit concentration quenching effects which are detrimental to light generation. The emitters are doped into host materials with a suitable weight percentage (wt.%) commonly between 1 wt.% and 20 wt.%. In this doped films, the transported charges recombine and excite the host molecules, then the energy is transported to the emitters. Thus, it is essential that the host materials possess high triplet states to realize this transfer effectively and smoothly. Additionally, when the emitter contains bulky dendritic peripheral groups, the concentration quenching can be weakened, allowing a host-free emissive layer structure. Also, the aggregation-induced emission (AIE) molecules can be used to realize a single material system or high doped percentage of guest materials, as the emission efficiency is in direct proportion to the aggregation concentration.[26-28]

Electron Transport Layer (ETL) / Hole Blocking Layer (HBL)

As the name already suggests, the electron transport layer (ETL) is used to facilitate the charge transport, similarly to the HTL. The difference is that the ETL is related to the electron transport. Electron deficient aromatic materials with matched LUMO levels such as triazoles, pyrimidines and pyridines are often used for this layer.[29] As this layer can also prevent the hole charge from leaving the emissive layer, it acts as a hole blocking layer (HBL).

Electron Injection Layer (EIL) and Cathode

An effective electron injection from the cathode is realized through an electron injection layer (EIL), similarly to the HIL and the anode. Mixtures of phenanthroline-based compounds or quinodimethane derivatives are often applied in this layer.[30] For the cathode, common base metals such as magnesium, aluminum or alloys of lithium are used.

Out-Coupling of Light

Lastly, for the usefulness of an electroluminescent device the EQE is a critical measure and is dependent on the device architecture and the optical properties of each layer. In addition, the cathode layer is opaque and out-coupling can only occur through the transparent anode. Before emission from the device, the photons have to pass several layers starting from the emissive layer, such as hole transport layer and hole injection layer, which inevitably cause refraction or trapping in waveguides. To optimize the EQE, optically beneficial micro-structuring of layers is a subsequent field of device research.[31]

1.1.2 Fabrication Techniques for OLEDs

There are two fundamental fabrication techniques which enable the construction of such layered devices on the microscopic scale, namely vacuum deposition and solution processing. In addition, a combination of these methods can be employed. Furthermore, the process of electroluminescence is particularly sensitive to impurities and therefore uppermost purity standards must be met both for the materials and solvents used.

1.1.2.1   Vacuum-deposited OLEDs

In 1987, the first OLED device was fabricated by Tang and Van Slyke *via* vapor deposition techniques.[2] Under high-vacuum, low molecular weight materials were vapored and deposited on the substrate to form different functional layers. This slow and sequential deposition process can smoothen and flatten the interfaces between neighboring layers, which favors the final light out-coupling. Due to its advantages, particularly with regards to purity as all materials are processed with the aid of vacuum and heat, this fabrication technique is widely used to produce industry-level displays for commercial applications. Though the vacuum deposition method is the industry standard to realize high EQE and long lifetimes, there are some disadvantages that limit its applications. For example, high molecular weight molecules or polymers cannot be sublimed for vacuum processing. Another disadvantage is the high cost to run and maintain the high-vacuum deposition chambers, particularly for large-size devices.

1.1.2.2 Solution-processed OLEDs

To circumvent these issues, solution processing has attracted intense attention and is explored because of the simple and cost-effective fabrication process and the possibility to use non-sublimable emitters such as polymers, and the ability of micro-structuring on surface.[32-34] The challenges associated with this processing technique are the orthogonality of solvents in order to avoid dissolution of underlying layers, the quantitative evaporation of solvent molecules and the uniformity of layers.

1.1.3 Emitters for OLEDs

In the electroluminescence process, singlet excitons and triplet excitons are generated in a 25% / 75% ratio from the recombination of electrons and holes based on Fermi-Dirac statistics (Scheme 1). Then, the singlet excitons relax from the excited state to the group state through the radiative deactivation process (fluorescent emission), and the similar process from the excited triplet states to the group states is called phosphorescent emission.

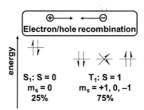

Scheme 1. Singlet and triplet excitons generated from the recombination of charge carriers. Image taken from Spuling.[18]

As shown in Scheme 2 (left), for most organic molecules, the transition from excited states to ground states is only allowed between two electronic states with the same spin multiplicity in theory, which means the transition between the excited singlet state (S_1) and the singlet ground state (S_0) is allowed (fluorescence), but it is forbidden from the excited triplet state (T_1) to S_0 (phosphorescence). Therefore, for the first generation of OLEDs, which uses purely fluorescent molecules as emitters, the internal quantum efficiency can only reach a maximum limit of 25%, which is the biggest limitation of potential for application. In addition, the large percentage of trapped T_1 states can relax vibronically or introduce degradation processes, limiting the device lifetime.[35]

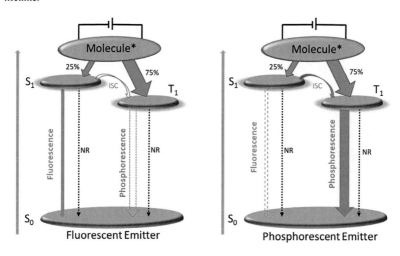

Scheme 2. Jablonski diagrams for purely fluorescent emitter (left) and phosphorescent emitter (right). The excited molecule by the applied electric field is indicated with *.

To obtain the remaining 75% of excitons in triplet states, phosphorescent heavy-metal complexes as emitters have been thoroughly studied and developed for the second generation of OLEDs. Through weakening the selection rules by an enhanced spin-orbit coupling induced by heavy metal atoms such as iridium or platinum, the resulting complexes are able to facilitate the internal system crossing (ISC, $S_1 \rightarrow T_1$) and accelerate the radiative deactivation from $T_1 \rightarrow S_0$ (phosphorescence) (Scheme 2, right).[36-43] With this novel triplet-harvesting strategy and relaxation pathway, the internal quantum efficiency of phosphorescent heavy-metal emitters can reach 100%, and eliminate triplet-based degradation. Though phosphorescent heavy-metal complexes as emitters have satisfied the basic requirements of commercial electroluminescent OLED devices in colors, device lifetimes and pipeline productions, the need of heavy and precious metals causes environmental concerns and cost issues.[44-45] Thus, eliminating the need of high-cost and environmentally precarious elements, while maintaining a complete theoretical limit for the internal quantum efficiency has been the aim of efforts in the last years. In this regard, the new and highly promising mechanism of thermally activated delayed fluorescence (TADF) has been developed and is being investigated and utilized with increasing interest both in academia and industrial research. The mechanism and synthetic approach to TADF emitters will be deeply described in the next chapter.

1.2   Thermally Activated Delayed Fluorescence (TADF)

The first example of TADF emitter based on organic materials was reported by Parker *et al.* in 1961, when they detected a delayed fluorescence component from Eosin (**1**) (Figure 3) at 70 °C in ethanol and initially named this phenomenon E-type delayed fluorescence.[46] This delayed fluorescence effect was also observed in benzophenone,[47] thioketones[48] and 9,10-anthraquinone.[49] In 1996, the rate equations to describe the time-resolved processes of the TADF was established by Berberan-Santos *et al.* in order to explain the emission mechanism of fullerene derivatives (**2**).[50] Though a lot of studies were devoted to utilizing purely organic molecules as TADF emitters for OLED devices, the early TADF OLEDs were still based on organometallic complexes using copper[51] or tin. In 2009, Adachi and co-workers reported the first OLED device based on a TADF molecule, tin(IV)-porphyrin complex (**3**).[52] Afterwards, Peters *et al.* fabricated a TADF OLED using bis(phosphino)diarylamido dinuclear copper(I) complex (**4**) as emitter and reaching an amazing EQE of 16.1% in 2010.[53]

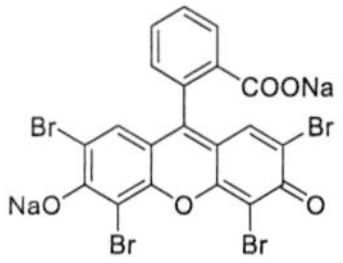

1
Eosin
1961

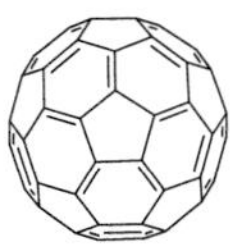

2
Fullerene
1997

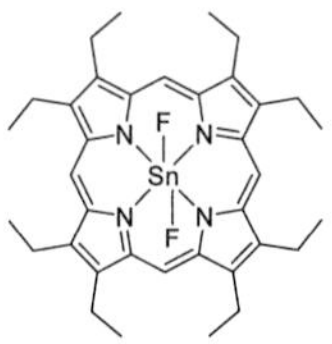

3
Sn(IV)-porphyrin
2009

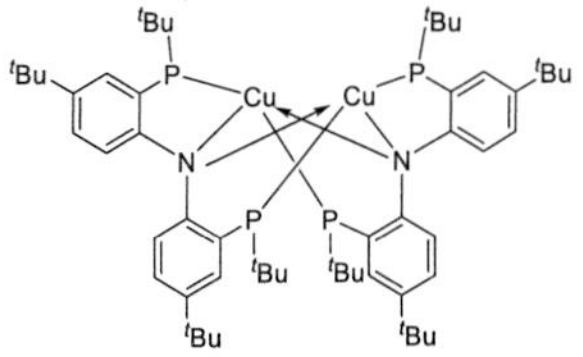

4
Cu(I) complex
2010

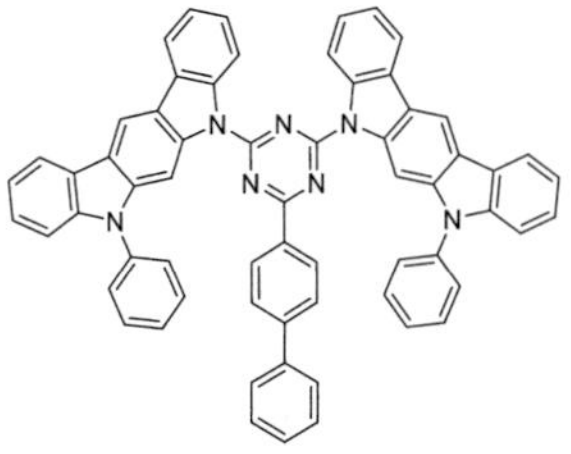

5
PIC-TRZ
2011

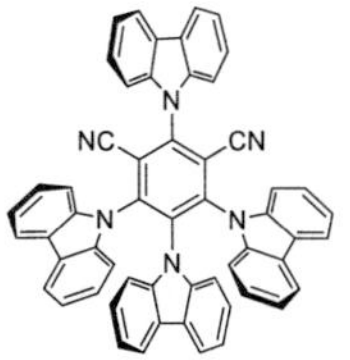

6
4CzIPN
2012

Figure 3. The development outline of TADF molecules.

The first TADF OLED using a purely organic molecule as emitter was reported by Adachi *et al.* in 2011.[54] **PIC-TRZ** (**5**) was employed as the emitter to construct an OLED device with an EQE of 5.3 %, approaching the theoretical limitation of conventional fluorescent materials. Combining the experimental data and the quantum mechanical analysis, they found that the reduction in the overlap of the highest occupied molecular orbital (HOMO) and the lowest unoccupied molecular orbital (LUMO) decreases the energy gap (ΔE_{ST}) between S_1 and T_1, leading to a reversibility of intersystem crossing needed for TADF emission. They also described that a twisted donor and acceptor structure induced by steric hindrance can effectively reduce the overlap between HOMO and LUMO. The real breakthrough was reported by the same group in 2012, when they designed and synthesized a series of TADF molecules derived from carbazolyl dicyanobenzenes with multiple carbazolyl units as donor groups and benzonitriles as acceptor groups.[55] Employing **4CzIPN** (**6**) as emitter in an OLED, an excellent device performance was achieved with an outstanding EQE of 19.3%. Compared to OLEDs based on the conventional fluorescence emitters with a maximum EQE of 5%, these astonishing results clearly demonstrate that the emission process comprises both singlet and triplet excitons through a highly efficient TADF mechanism.

Moreover, as the device performance of purely organic TADF OLEDs became comparable with the best phosphorescent OLEDs based on heavy and noble metal atoms, purely organic TADF emitters (3rd generation emitters) have now succeeded phosphorescent (2nd generation) emitters as the ideal candidates for OLED applications.

1.2.1 Working Principle of TADF

After the milestone work by Adachi and co-workers, TADF has attracted tremendous attention in the organic optoelectronic materials field and has since been considered as the most promising approach for OLED applications.[56-59] The principle of the TADF mechanism is still under continuous investigation and improvement in order to guide and accelerate molecular design. As shown in Scheme 3 (left), when the energy gap (ΔE_{ST}) between the S_1 and T_1 is minute and the lifetime of the T_1 excitons is long enough, the formally spin-forbidden reverse intersystem crossing (RISC) can be thermally activated and the triplet excitons can be up-converted to singlet excitons. The original singlet excitons and the newly generated singlet excitons relax to the ground state through prompt and delayed fluorescence emission, respectively, resulting in 100% internal quantum efficiency.

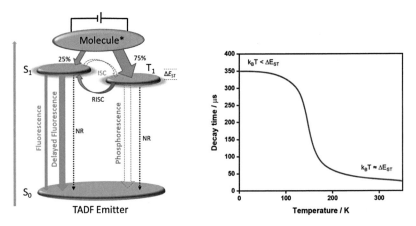

Scheme 3. Jablonski diagram of TADF emitters (left). Temperature dependence of the emission decay time (right).

Therefore, for an ideal TADF molecule, the RISC up-conversion is the key process and the rate constant of RISC (k_{RISC}) is affected by ΔE_{ST} according to the relationship between them, described by the Boltzmann distribution:

$$k_{RISC} \propto \exp\left(-\frac{\Delta E_{ST}}{k_B T}\right) \qquad (1)$$

where k_B is the Boltzmann constant and T is the temperature.[57] As shown in Equation 1, the ratio of ΔE_{ST} and the thermally available energy $k_B T$ must be low in order to promote RISC. As the thermal energy ($k_B T$ at 298K equals 0.026 eV) is fixed if the working conditions of OLEDs are expected to be at room temperature, a small ΔE_{ST} must be obtained through careful molecular design. For most conventional fluorescent molecules, the RISC process is impossible to activate at room temperature due to the large ΔE_{ST} (0.5–1.0 eV). In addition, a large ΔE_{ST} often leads to a longer delayed fluorescence lifetime τ_d, which can be decreased with an increasing rate of ISC (k_{ISC}) or RISC (k_{RISC}).[60-62] The relationship between them can be expressed through:[63]

$$\frac{1}{\tau_d} = k_{nr}^T + \left(1 - \frac{k_{ISC}}{k_r^S + k_{nr}^S + k_{ISC}}\right) k_{RISC} \qquad (2)$$

where the radiative and nonradiative decay rate constants of the singlet state are represented by k_r^S and k_{nr}^S.

Combining these analyses, ΔE_{ST} (the energy gap between S_1 and T_1) is considered as one of the decisive factors to the TADF process. Normally, ΔE_{ST} is proportional to the exchange integral J, which depends on the overlap between HOMO and LUMO, when the excited singlet states (S_1) and triplet states (T_1) are assumed to be dominated by the transition from HOMO to LUMO. The relationship between ΔE_{ST} and J, and the HOMO/LUMO orbital overlap can be expressed through Equation 3 and Equation 4:[64]

$$\Delta E_{ST} = E_S - E_T = 2J \qquad (3)$$

$$J = \iint \phi_{HOMO}(r_1)\phi_{LUMO}(r_2)\frac{1}{|r_2 - r_1|}\phi_{HOMO}(r_2)\phi_{LUMO}(r_1)\,dr_1\,dr_2 \qquad (4)$$

where ϕ_{HOMO} and ϕ_{LUMO} are the spatial distributions of the HOMO and the LUMO, and r_1 and r_2 are position vectors.[65] Based on these descriptions, the trend is that decreasing the overlap between HOMO and LUMO leads to a smaller exchange integral J (Equation 4) and a following smaller ΔE_{ST} (Equation 3). Therefore, the strategy of HOMO and LUMO separation is an effective method to create TADF emitters.

Besides the small ΔE_{ST}, a large radiative rate constant (k_r) of the transition from S_1 to S_0 is also critical to obtain highly efficient TADF emission. However, the relationship between a large k_r and a small ΔE_{ST} is conflicting,[56] so how to balance the correlation between a small ΔE_{ST} and large k_r must be fully and deeply studied.

1.2.2 Molecular Design Strategies for TADF Emitters

Vast and continuous attention and efforts have been devoted to obtaining highly efficient TADF emitters, and the working principles and molecular design strategies have also been carefully studied and stated. Typically, for purely organic TADF molecules, a small ΔE_{ST} activates RISC from the excited triplet state to the excited singlet state, and therefore plays the key goal in order to enable TADF emission. A minute ΔE_{ST} is obtained through twisted or weakly-conjugated donor-acceptor systems resulting in a reduced overlap between HOMO and LUMO. Then, through charge transfer transitions, TADF emission is realized. In addition, multi-resonance effects can also be used to separate HOMO and LUMO, providing an alternative design approach for TADF molecules.

1.2.2.1 Donor-Acceptor Induced Charge Transfer

Using twisted donor-acceptor structures to separate HOMO and LUMO was first reported by Adachi and co-workers and it has been proven to be a reliable and successful method to construct TADF molecules. As TADF donor groups, electron-donating units, such as carbazole-based groups,[66] triphenylamines[67] or acridine-based structures like phenoxazines,[68] phenothiazine[69] and 9,9-dimethyl-9,10-dihydroacridine[70] are commonly employed (Figure 4). Electron-accepting units, like benzonitriles,[61] triazines,[71] benzophenones,[72] sulfones[73] and dimesitylboranes,[74] are predominantly used as acceptors. According to the pathway of charge transfer, these can be classified into three types.

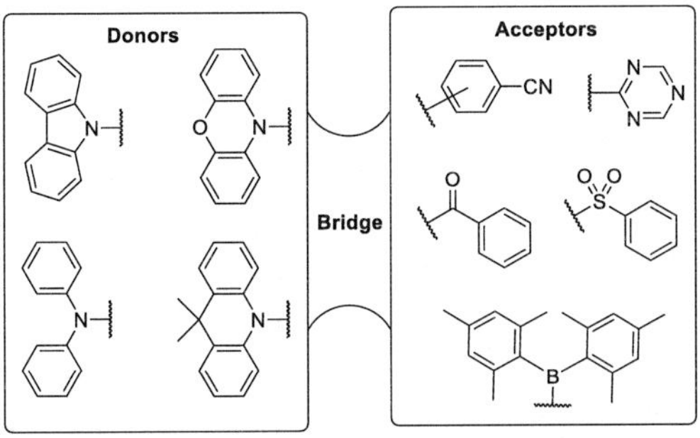

Figure 4. Typical donors and acceptors for donor-acceptor TADF emitters.

1.2.2.1.1   Torsion Induced Charge Transfer

In 2011, Adachi and co-workers reported a purely organic TADF emitter **4CzIPN (6)**,[55] which was constructed by a twisted donor-acceptor structure of an isophthalonitrile acceptor unit decorated with four carbazole donor units leading to a very crowded and therefore twisted arrangement and then a limited overlap of HOMO and LUMO (Figure 5). They found that the increased torsion between donor and acceptor led to the reduction of spatial overlap between HOMO and LUMO, resulting in a very small ΔE_{ST} (0.08 eV). Evaluation of this TADF molecule as emitter showed a very high photoluminescence efficiency in excess of 90%, and an external device efficiency of more than 19%, which fully proved that the charge transfer between the twist donor and acceptor-based emitter was highly efficient. After this seminal work, the introduction of steric hindrance by bulky substituents to twist donor and acceptor became the most popular design strategy to create TADF molecules. When one more carbazole group was introduced to the central cyanobenzene unit, a new emitter **5CzBN (7)** with five carbazole groups was constructed (Figure 5).[75] Photophysical studies demonstrated that it was also an excellent TADF emitter.

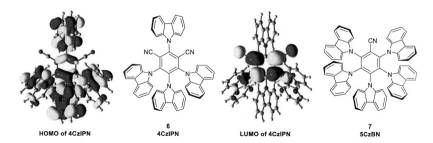

| HOMO of 4CzIPN | 6
4CzIPN | LUMO of 4CzIPN | 7
5CzBN |

Figure 5. HOMO and LUMO of **4CzIPN** (**6**) (copyright: Nature) and chemical structure of **5CzBN** (**7**).

With regards to the twisted donor-acceptor approach compound **CzPhTrz (8)** has become an important comparison structure, as this emitter itself does not show TADF due to the large ΔE_{ST} of 0.30 eV and only limited twist between the donor and acceptor unit, but a deep blue emission. In 2015, Wu *et al.* reported the derived TADF emitter **DMAC-TRZ (9)** using 9,9-dimethyl-9,10-dihydroacridine as electron-donating moiety to replace the carbazole group of *N*-carbazolyl-4-triphenyltriazine **CzPhTrz (8)**.[70] In contrast to **CzPhTrz (8)**, the dihedral angle between donor and acceptor of **DMAC-TRZ (9)** is increased from 51.5° to 88.0° induced by the closer proximity of the adjacent hydrogen atoms, leading to a ΔE_{ST} of 0.05 eV. A very high PLQY of 90% is achieved in doped film. Using **DMAC-TRZ (9)** as TADF emitter, excellent OLED performances with EQEs of 26.5% and 20.0% are obtained for doped and nondoped devices, respectively.

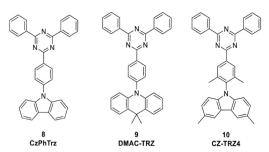

| 8
CzPhTrz | 9
DMAC-TRZ | 10
CZ-TRZ4 |

Figure 6. Chemical structures of **CzPhTrz (8)**, **DMAC-TRZ (9)** and **CZ-TRZ4 (10)**.

Based on **CzPhTrz (8)**, Adachi *et al.* introduced methyl groups on the linking 1,4-substituted phenyl unit to increase the steric hindrance between the carbazole group and the bridging phenyl unit.[76] The dihedral angle between donor and acceptor of the resulting **CZ-TRZ4 (10)** emitter is increased to 82.3 and a small ΔE_{ST} of 0.15 eV is realized, which again activated the RISC process, resulting in an efficient TADF emitter with a high PLQY of 85%. It is worthy to be stressed that OLED devices based on **CZ-TRZ4 (10)** exhibit a high EQE of 18.3% and deep blue emission with CIE coordinates of (0.150, 0.097).

Moreover, Swager *et al.* reported that the TADF process could be activated using a donor with a large and spherical conjugation structure. A triptycenylcarbazole group was used as donor to create emitters **TCZ-TRZ (11)** and **TCZ-TRZ(Me) (12)** by replacing the carbazole group of the parent molecule **CzPhTrz (8)** (Figure 7).[77] Without additional torsion introduced, the dihedral angle between donor and acceptor for emitter **TCZ-TRZ (11)** is 51.6°, which is identical to **CzPhTrz (8)** (51.5°), but the ΔE_{ST} of **TCZ-TRZ (11)** is decreased to 0.27 eV due to the dissipation of the HOMO orbital on the extended triptycenyl unit, reducing the HOMO/LUMO overlap and activating the TADF emission. In order to further decrease ΔE_{ST}, they introduced one methyl group on the bridging phenyl ring to increase steric hindrance between donor and acceptor leading to compound **TCZ-TRZ(Me) (12)**. The ΔE_{ST} was effectively reduced from 0.27 eV to 0.16 eV with an improved k_{RISC} from 6.0×10^2 s^{-1} to 35.3×10^2 s^{-1}. Furthermore, Spuling *et. al.* reported that TADF emission could be turned-on *via* the torsion generated from the chiral [2.2]-paracyclophanyl scaffold on the carbazole donor leading to TADF emitters **(13)**.[18, 78]

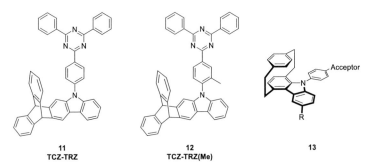

11
TCZ-TRZ

12
TCZ-TRZ(Me)

13

Figure 7. Chemical structures of **TCZ-TRZ (11)**, **TCZ-TRZ(Me) (12)** and [2.2]paracyclophane-based TADF emitters (**13**).

1.2.2.1.2   Through-Space Induced Charge Transfer

Introducing torsion between donor and acceptor has become the major strategy to design TADF emitters, as it is a very effective approach to decrease the ΔE_{ST} of an organic molecule. However, as mentioned above, a large dihedral angle suppresses the radiative decay from S_1 to S_0 and results in reduced internal quantum efficiency. In 2015, Cheng and co-workers designed and synthesized two molecules, **DCBPy (14)** and **DTCBPy (15)**, based on benzoylpyridine-carbazole (Figure 8).[79] Very small ΔE_{ST} values of 0.03 (**DCBPy, 14**) and 0.04 eV (**DTCBPy, 15**) are obtained and the photophysical properties demonstrate that these two emitters show highly efficient TADF processes with photoluminescent quantum yields up to 88.0% (**DCBPy, 14**) and 91.4% (**DTCBPy, 15**), respectively, when doped in thin films. The crystal structure of **DTCBPy (15)** suggests that strong intramolecular space interaction exists between the *ortho tert*-butyl carbazole unit (donor) and the pyridine group (acceptor) facilitated by the short distance of 2.9−3.7 Å between them. They explained that the through-space interactions likely play a key role to obtain very small ΔE_{ST} and high quantum efficiency simultaneously.

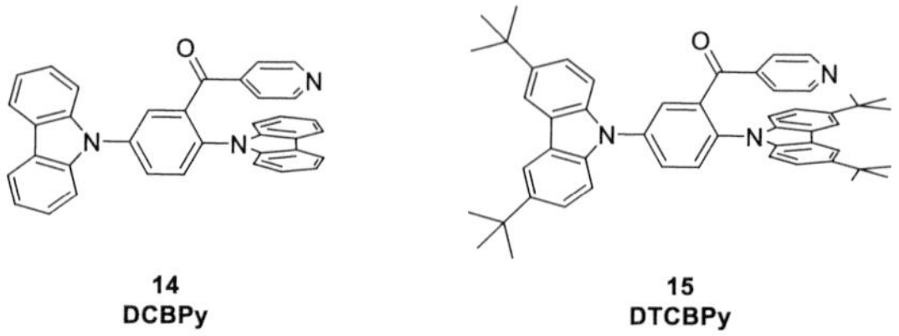

14
DCBPy

15
DTCBPy

Figure 8. Chemical structures of **DCBPy (14)** and **DTCBPy (15)**.

Swager and co-workers reported three molecules bearing donor and acceptor units bridged by 9,9-dimethylxanthene (Figure 9).[80] Computations predicted that these molecules possess very small ΔE_{ST} values in the range of 0.001–0.007 eV and low oscillator strengths of 0.00007–0.0005. The very small ΔE_{ST} values suggested RISC from T_1 to S_1 could be efficiently facilitated. However, the very weak oscillator strengths also indicated that the charge transfer between HOMO and LUMO was almost suppressed, resulting in low emission efficiency. Indeed, the quantum yields of these three emitters were very low (1.0~2.1%) in solution. When oxygen was removed from the solution, the yield could be partly enhanced (5.9~7.7%) and the lifetimes in the region of microsecond were obtained, which indicates a delayed emission component, such as in TADF emitters. Further, increased quantum yields of 35~66% were observed in the solid state. Crystal structure analyses demonstrated that through-space π–π interactions between the donor and acceptor exist, however, the intermolecular communications through C–H$\cdots\pi$ interactions is the main pathway to dominate the light emission, resulting in typical aggregation-induced emission.

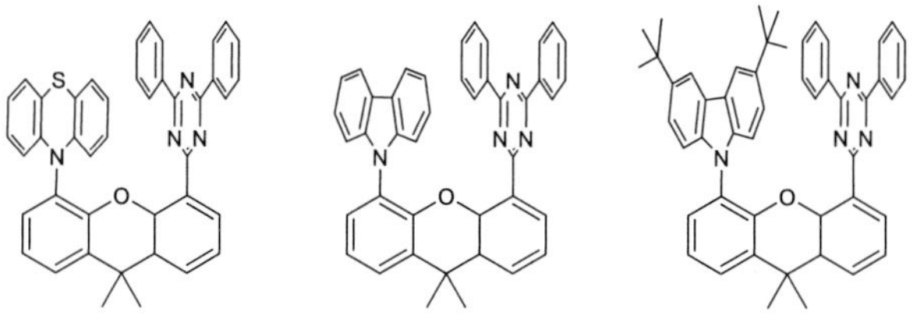

16	**17**	**18**
XPT	**XCT**	**XtBuCT**
ΔE_{ST} = 0.0001 eV	ΔE_{ST} = 0.008 eV	ΔE_{ST} = 0.007 eV
f = 0.00007	f = 0.0003	f = 0.0005
λ_{EL} = 584 nm	λ_{PL} = 418 nm	λ_{EL} = 488 nm
EQE_{max} = 10%		EQE_{max} = 4%

Figure 9. Chemical structures and photophysical properties of **XPT (16)**, **XCT (17)** and **XtBuCT (18)**.

Another intriguing through-space approach is based on the [2.2]paracyclophane scaffold. The distance of 3.09 Å between two bent phenyl rings of [2.2]paracyclophane (PCP) is shorter than the van der Waals distance between layers of graphite (3.35 Å),[81] thus the electronic communication in a single molecule can be realized smoothly *via* through-space charge transfer.[82-86] Based on these studies, Bräse *et al.* reported two TADF molecules *via* combining the donor and acceptor units on the different decks of [2.2]paracyclophane (PCP) (Figure 10). Moderate quantum yields of 45% (***trans*-Bz-PCP-TPA, 19**) and 60% (***cis*-Bz-PCP-TPA, 20**) in deoxygenated solution and delayed lifetimes of 1.8 μs (***trans*-Bz-PCP-TPA, 19**) and 3.6 μs (***cis*-Bz-PCP-TPA, 20**) were achieved.[87] Moreover, Zhao and co-workers also reported two TADF molecules (***m*-BNMe2-Cp, 21** and ***g*-BNMe2-Cp, 22**) based on [2.2]paracyclophane (PCP) by employing dimethylamino as electron-donating units and dimesitylboryl group (BMes2) as electron-accepting parts (Figure 10).[88] TADF properties of ***g*-BNMe2-Cp (22)** with photoluminescence quantum yields up to 72% in solution and 53% in solid state were obtained by through-space induced charge transfer process.

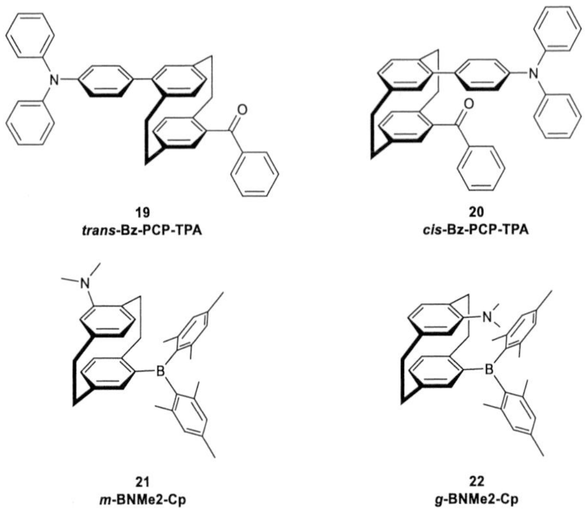

Figure 10. Chemical structures of *trans*-Bz-PCP-TPA (**19**), *cis*-Bz-PCP-TPA (**20**), *m*-BNMe₂-Cp (**21**) and *g*-BNMe₂-Cp (**22**).

1.2.2.1.3   Intermolecular Interaction Induced Charge Transfer

Recently, Zhang and co-workers reported a non-doped TADF OLED device with a novel organic molecule as emitter. The molecule **DMAC-*o*-TRZ** (**23**) contains the donor (1,10-phenyl-9,9-dimethyl-9,10-dihydroacridine) and the acceptor (triphenyltriazine) which are connected by a spacious and conjugation-forbidden linkage (oxydibenzene) (Figure 11).[89] A host-guest OLED device employing **DMAC-*o*-TRZ** (**23**) as emitter was constructed, showing a bad device performance with an EQE of only 3.0%. Interestingly, with increasing of doping concentration, the device performances were significantly improved. When the doped concentration was 42%, a high EQE of 15.5% was obtained with deep blue emission. The doped concentration was gradually increased to 100% (without host material), while the EQE still maintained 14.7% for this non-doped device. Due to the almost fully broken conjugation between donor and acceptor, the intramolecular charge transition was almost completely suppressed. With the increase of the doping concentration, intermolecular charge transfer transition was enabled and acted as the main relaxation pathway in this emitter. The simple fabrication procedure as a neat film and excellent device performances demonstrate that it is a promising strategy for non-doped TADF OLEDs to employ emitters which allow highly efficient intermolecular charge transfer transition.

Figure 11. Chemical structure of **DMAC-*o*-TRZ (23)** (left) and the process of intermolecular interaction (right) (copyright: Wiley-VCH).

1.2.2.2   Multi-Resonance Effect

In 2016, Hatakeyama and co-workers established a novel molecular design strategy to construct TADF emitters **DABNA-1 (24)** and **DABNA-2 (25)** that showed pure blue and very narrow emission with a full width at half maximum (FWHM) of only 28 nm (Figure 12).[90] Through the opposite resonance effect of nitrogen atoms and boron atoms, the spatial densities of frontier molecular orbitals (FMOs) were separated effectively without the need of conventional donor-acceptor structures. The LUMO is located on the central boron atom and the *ortho* and *para* carbon positions on the phenyl rings relative to it, while the HOMO is mainly localized on the nitrogen atoms and *ortho* and *para* carbon positions on the phenyl rings relative to them (Figure 12b), which ingeniously permit the separation of HOMO and LUMO without the need of spatial separation of donor and acceptor groups. Outstanding optoelectronic properties were obtained with blue emission at 459 nm, CIE coordinates of (0.13, 0.09) and a high EQE of 13.5% for **DABNA-1 (24)**. Moreover, the oscillator strength could be increased from 0.205 to 0.415 when substituents were introduced to the parent molecule (**DABNA-1, 24**). Using **DABNA-2 (25)** as emitter, a TADF OLED was fabricated with excellent device performances with blue emission at 467 nm, CIE coordinates of (0.12, 0.13) and a further increased the maximum EQE of 20.2%. It is worth mentioning that these two emitters both possess very thin FWHM of only 28 nm, resulting in more pure color emission than conventional donor-acceptor TADF emitters.

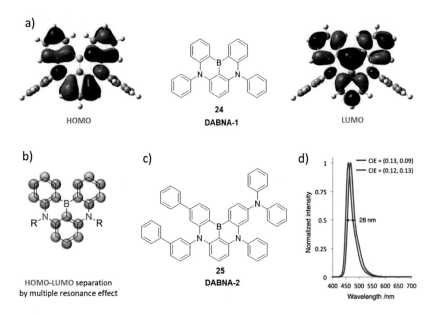

Figure 12. a) Chemical structure of **DABNA-1** (**24**) and HOMO (left) and LUMO (right) of it. b) HOMO-LUMO separation by multiple resonance effect. c) Chemical structure of **DABNA-2** (**25**). d) Electroluminescence spectra of **DABNA-1** (**24**, blue line) and **DABNA-2** (**25**, red line) (copyright: Wiley-VCH).

Enlighted by the above results, Jiang and co-workers designed and synthesized two new TADF emitters, **QAD** (**26**), **QAD-DAd** (**27**), based on the multi-resonance effect using a fused amine/carbonyl system (Figure 13). Employing **QAD (26)** as emitter, pure blue OLEDs with EQE of 19.4% and a small full width at half maximum of 39 nm were accessed.[91] In addition, QAD-based unit could be used as acceptor to construct new TADF emitter (**QAD-Dad, 27**), resulting in highly efficient OLEDs with EQE of 23.9% and 19.3% for vacuum-deposited and solution-processed devices, respectively. Furthermore, through the introduction of mesityl groups to reduce the aggregation-caused quenching, the PLQY of 80% and device performance with EQE of 21.1% were realized by using **Mes₃DiKTa** (**28**) as emitter reported by the group of Zysman-Colman.[92]

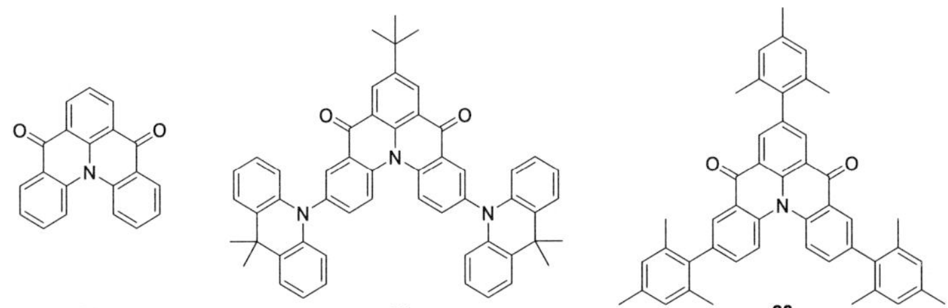

Figure 13. Chemical structures of **QAD (26)**, **QAD-DAd (27)** and **Mes₃DiKTa (28)**.

1.3 Molecular Design Towards Efficient TADF OLEDs

As described previously, the established strategy in order to obtain OLEDs with satisfying EQEs is to start by designing, synthesizing and selecting emitters which show high IQEs. A lot of studies are devoted to exploring new donors, acceptors and molecular linking types between them to reach this goal, and many TADF emitters with IQEs which are very close to 100% are reported now.[93-95] Another parameter to optimize the EQE of device is the architecture of the OLED stack. In this regard the out-coupling coefficient of the generated photons is a critical factor in order to reach excellent device performances.[96-100] It is demonstrated that molecules with specific orientations in the emissive layer can improve the intrinsic out-coupling factor, leading to enhanced device performances.

1.3.1 High Internal Quantum Efficiency Emitters for TADF OLEDs

According to the method to obtain high IQEs and EQEs, OLEDs can be classified into two types, host-guest OLEDs (doped OLEDs) and host-free OLEDs (non-doped OLEDs).

1.3.1.1 Emitters for Host-Guest TADF OLEDs

In the emissive layer of the OLED stacks, the concentration quenching is a common phenomenon for organic materials due to the molecular stacking or exciton-exciton quenching. In order to solve this problem, doping the emissive molecules in a host material is an efficient way to suppress concentration quenching. This is a classic strategy to obtain high IQEs and thus highly efficient

OLEDs, which is suitable for most TADF emitters. It is worth mentioning that rigid and bulky emitters can retrain the molecular stacking and hence lead to a weakened intermolecular quenching even at relativly high doping concentrations.[101]

1.3.1.2   Emitters for Host-free TADF OLEDs

Though the host-guest system is the predominant fabrication method for TADF OLEDs, the manufacturing process of devices becomes more complicated, especially for vacuum-deposited devices, as more than one material need to be deposited at the same time which requires careful control of the evaporation speed and ratio of each material. In order to reduce cost and complexity, host-free systems, solely employing the emitter for the emissive layer, are considered as a reasonable fabrication strategy. For the molecular design strategies, it has been demonstrated that molecules which possess bulky, isolating and electronically innocent peripheral units, large twisted conformation and aggregation-induced emission properties are the promising candidates for highly efficient non-doped OLEDs.

Adachi and co-workers reported two TADF emitters, **DMAC-DPS (29)** and **DMAC-BP (30)**, with high PLQYs of 88% for **DMAC-DPS (29)** and 85% for **DMAC-BP (30)** in neat films (Figure 14). Highly efficient non-doped OLEDs were achieved with a maximum EQE of 19.5% and 18.9% using these two molecules as emitters.[102] In particular, the EQE still retained at 18.0% for **DMAC-BP (30)** at a practical luminance of 1000 cd m^{-2}, which can be attributed to its shorter delayed lifetime. In 2018, Yasuda *et al.* demonstrated very high PLQY (close to 100%), narrow emission bandwidths, short emission lifetimes of ≈1 μs, and a fast reverse intersystem crossing rate of over 10^6 s^{-1} based on the TADF emitters derived from dibenzoheteraborins.[103] Employing these molecules as emitters, excellent non-doped blue OLEDs were achieved with CIE coordinates of (0.15, 0.36) and a maximum EQE of 22.8% for **MPAc-BS (31)** and CIE coordinates of (0.14, 0.23) and a maximum EQE of 21.3% for **MPAc-BN (32)** because of the high PLQYs and fast k_{RISC} of the emitters (Figure 14).

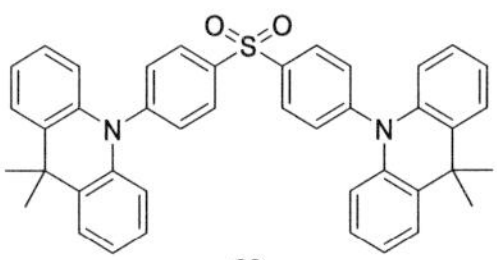

Figure 14. Chemical structures of **DMAC-DPS (29)**, **DMAC-BP (30)**, **MPAc-BS (31)** and **MPAc-BN (32)**.

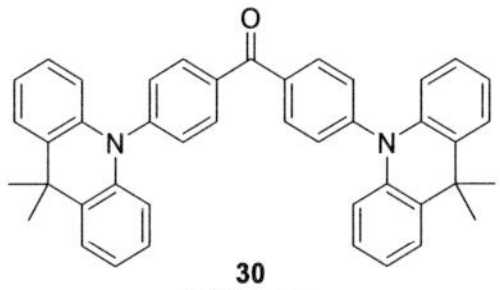

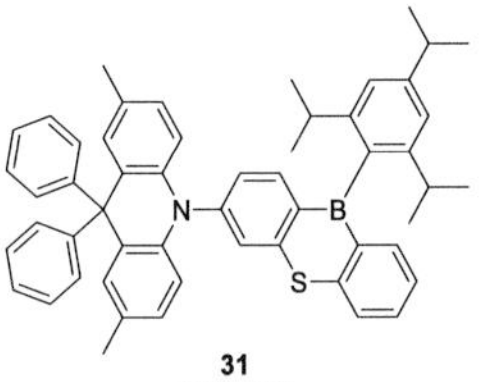

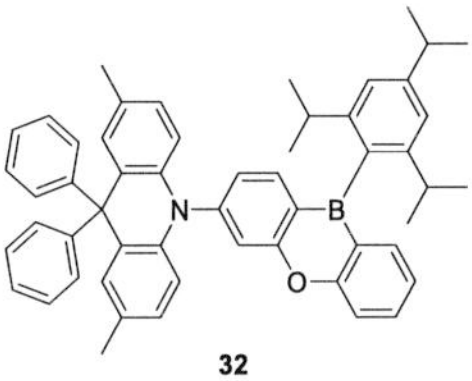

1.3.2 Orientation Controlled Emitters for TADF OLEDs

Recent studies show that the transition dipole moment of some TADF molecules exhibits highly horizontal orientation to the plane of substrate in the deposited films. Since the direction of light emission is generally perpendicular to the transition dipole moment of the emitter, the ratio of all the light generated by the emitter molecules that can pass through the substrate also becomes higher, leading to an increased out-coupling factor. For the molecular construction strategies, linear or planar molecular structures are considered to be promising candidates to produce preferential horizontal orientation transition dipole moments.[104-107] Combining the requirement of donor-acceptor TADF emitters, molecules constructed of linear structure are confirmed to be reasonable and preferable, because the energy states of them can be conveniently controlled and tuned, which is challenging to achieve for planar molecules. Additionally, for most linear molecules, the orientation of transition dipole moment is in the line with its long axis. Therefore, the desired control of orientation of the transition dipole moment can be realized through careful device manufacturing or synthetic derivatization of linear donor-acceptor TADF emitters.

In 2015, Adachi and co-workers reported the donor-acceptor molecule **DACT-II (33)** bearing a triazine core as acceptor and a carbazole enlarged with two aryl amines as donor, which could effectively reduce the ΔE_{ST} to a minute 0.009 eV, resulting in almost 100% photoluminescent yield.[108] Excellent green TADF device performances with CIE coordinates of (0.21, 0.50) and

EQE of 29.6% were achieved by using **DACT-II (33)** as emitter, which exhibited highly horizontal orientation in solid film. When the donor was changed to the spiroacridine by Wu and co-workers, the linear shaped molecule **SpiroAC-TRZ (34)** was obtained and showed preferentially horizontally oriented transition dipoles of 83%.[109] Benefiting from the high degree of horizontal orientation, a green TADF OLED with an EQE of 36.7% was obtained. Furthermore, a donor derived from indolocarbazole was also explored to construct TADF emitters. Due to the pronounced linear shape, the OLED based on **IndCzpTr-2 (35)** showed an EQE of 30.0% and green emission with CIE coordinates of (0.23, 0.50).[110] Recently, Su *et al.* designed and synthesized a TADF molecule *via* further linear modification of the spiroacridine donor unit. By introducing the nonconjugated fragment to prolong the length of molecule along the long axis, the horizontal orientation dipole ratio of **TspiroS-TRZ (36)** reaches up to 90%. Using this emitter, an efficiency breakthrough was achieved for sky-blue TADF OLEDs with a CIE coordinates of (0.17. 0.33) and an EQE of 33.3%.[111]

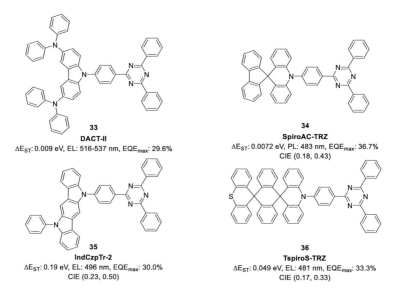

33
DACT-II
ΔE_{ST}: 0.009 eV, EL: 516-537 nm, EQE_{max}: 29.6%

34
SpiroAC-TRZ
ΔE_{ST}: 0.0072 eV, PL: 483 nm, EQE_{max}: 36.7%
CIE (0.18, 0.43)

35
IndCzpTr-2
ΔE_{ST}: 0.19 eV, EL: 496 nm, EQE_{max}: 30.0%
CIE (0.23, 0.50)

36
TspiroS-TRZ
ΔE_{ST}: 0.049 eV, EL: 481 nm, EQE_{max}: 33.3%
CIE (0.17, 0.33)

Figure 15. Chemical structures and performance of known linear triazine-based TADF emitters, **DACT-II (33)**, **SpiroAC-TRZ (34)**, **IndCzpTr-2 (35)** and **TspiroS-TRZ (36)**.

2 Objective

The development of TADF emitters is considered as one of the most promising approaches to expand the practical applications of OLEDs due to its energy saving, diminished environmental and cost concerns, and access to flexible and pure black displays.[56] The design principle of spatially separated donor-acceptor structures has been proven to be a reliable and successful method to construct TADF emitters. Electron-accepting units such as benzonitriles, triazines, benzophenones, sulfones and dimesitylboranes are used as acceptors because of their suitable electron-withdrawing ability and high emission efficiency. For donor groups, electron-donating units, like carbazole-based groups, triphenylamines or acridine-based structures are often employed.[58] Considering the variety and excellent performance of acceptors, a further exploration of novel donors in order to obtain and evaluate ideal TADF emitters is essential.

Therefore, the aim of this thesis is to design, synthesize and characterize TADF emitters based on N-heterocyclic donors for application in OLEDs, particularly blue OLEDs (Figure 16).

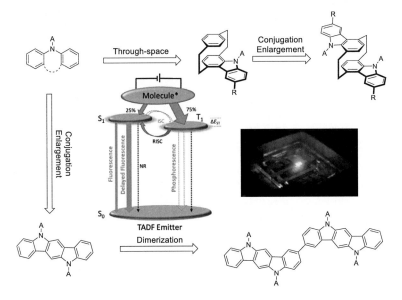

Figure 16. Objectives of this thesis.

In the first part, turn-on of TADF emission of boron acceptor-based emitters through variation of N-heterocyclic donors is investigated. DFT supported evaluation of a set of molecules is performed before the synthesis. Then, thorough experimental measurements are conducted to confirm and evaluate their TADF properties.

Secondly, the linear TADF emitters with the indolocarbazole donor are designed, synthesized and characterized. Their TADF properties and the role of peripheral *tert*-butyl groups on the triazine acceptor are evaluated with regards to the degree of preference for horizontal orientation. To expand the use of this unique donor for OLED applications, a synthetic method to introduce steric hindrance between donor and acceptor and to dimerize the derivatives is developed.

Lastly, a series of blue and deep-blue TADF emitters are designed and synthesized through introducing electron-withdrawing groups to the parent donor (**Czp**), or enlarging the conjugation through the synthesis of a through-space bis-conjugated donor **DCCP**. In order to obtain these emitters, a synthetic access on multigram scale to the donor units and methods to introduce functional groups are optimized. Furthermore, thorough optoelectronic characterization of these emitters is conducted to evaluate their suitability for OLED applications.

3 Results and Discussion

3.1 N-Heterocyclic Donors for Boron-Containing TADF Emitters

3.1.1 OBO-Based TADF Emitters

Studies show that three-coordinate boron units possess high triplet states and suitable electron-withdrawing abilities, which can be used as acceptors to create blue emitters. However, three-coordinate boron compounds suffer from poor chemical- and photo-stability due to the empty p_z-orbital rendering the isoelectronic situation to the carbonium ion, which can be attacked by nucleophiles, resulting in the formation of radical anions and polycyclic aromatic hydrocarbons.[112] In order to increase the stability of these compounds, steric hindrance is added in the vicinity of the boron atom to protect it. This leads to air-stable derivatives. Kaji and co-workers synthesized a number of stable boron compounds by using triarylboron-based (**37**) unit as acceptor and realized efficient TADF OLEDs with sky-blue emission.[74] In addition, positioning the boron atom at the centre of polycyclic π–systems is another strategy to achieve stable boron compounds. Kitamoto *et al.* reported a number of light-blue and green TADF emitters by using 10-*B*-phenoxaboryl groups as the electron-accepting unit (**38**).[113] Moreover, Hatakeyama and co-workers reported a stable OBO-fused benzo[fg]tetracene with high PLQY, large optical band gap and a suitably small radiative rate constant (**39**).[114] Additionally, the Müllen group also reported on similar OBO-based compounds (**40**), which exhibited good stability, strong fluorescence and a larger optical gap than that of the carbon analogue bistetracene.[115] Although this OBO-based acceptor exhibited very promising potential to be applied in optoelectronics, TADF emitters based on it have not been fully studied.[116]

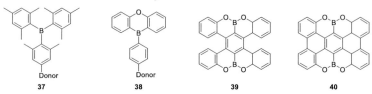

Figure 17. TADF emitters based on triarylboron unit (**37**) and 10-*B*-phenoxaboryl group (**38**). Chemical structures of **39** and **40**.

3.1.1.1 Molecular Design and DFT Calculation

Enlighted by the above molecular design strategies, two acceptor-donor-acceptor molecules were designed by employing OBO-fused benzo[fg]tetracene (OBO) as acceptor (Figure 18). Two *tert*-butyl groups were introduced to the acceptor to maintain solubility. For the donor part, 5,11-dihydroindolo[3,2-b]carbazole group was used as the central donor because of its large planar rigid conjugation structure, suitable electron-donating strength and excellent charge transporting ability.[110, 117] Moreover, the two antisymmetric reactive positions facilitate the construction of linear molecules, which is crucial to obtain horizontal orientation preference of the emitter in solid films. For **ICz-*t*BuOBO** (**41**), the donor group and acceptor units were connected directly. **ICz-Ph-*t*BuOBO** (**42**) was created by introducing two phenyl rings as bridges between OBO and central donor to beneficially modulate the magnitude of HOMO/LUMO interaction, which is in favor of leading to high photoluminescence quantum yield.[118] Additionally, molecule **Cz-*t*BuOBO** (**43**) based on the carbazole group was also designed for comparison.

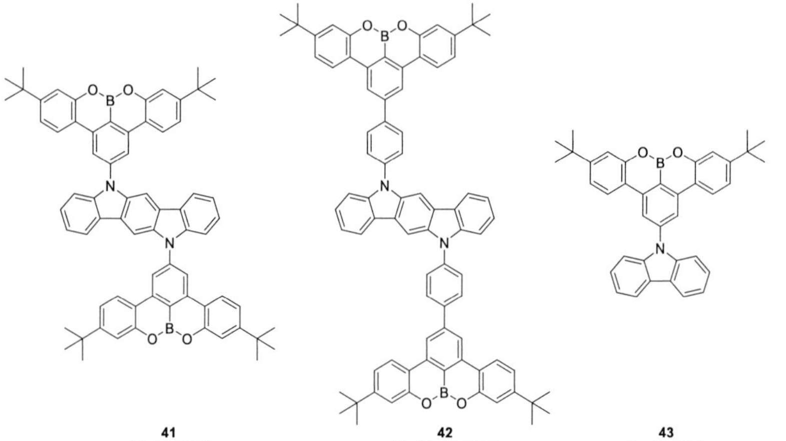

41 ICz-*t*BuOBO	**42** ICz-Ph-*t*BuOBO	**43** Cz-*t*BuOBO

Figure 18. Chemical structures of **ICz-*t*BuOBO** (**41**), **ICz-Ph-*t*BuOBO** (**42**) and **Cz-*t*BuOBO** (**43**).

Density functional theory (DFT) calculations were performed to assess the electronic structure of these molecules using the Gaussian 09 revision D.018 suite. The PBE0 functional with the standard Pople 6-31G (d,p) basis set was used to optimize the geometries of the molecular structures in the ground state in gas phase. Next, time-dependent DFT calculations were performed employing the

Tamm-Dancoff approximation (TDA) based on the ground state optimized molecular structures. The GaussView 5.0 software was employed to visualize the molecular orbitals.

DFT results show that the HOMO of **ICz-*t*BuOBO (41)** is mainly dispersed on indolocarbazole unit and partly on the central phenyl unit of OBO, while the LUMO is localized on the whole OBO moiety (Figure 19). As a result, the HOMO and LUMO overlap on the phenyl units. Compared to **Cz-*t*BuOBO (43)** (Figure 20), the ΔE_{ST} of **ICz-*t*BuOBO (41)** is decreased from 0.65 eV to 0.56 eV, but it is still too large to realize efficient RISC, which suggests that it is not a good candidate for TADF emission. For **ICz-Ph-*t*BuOBO (42)**, the LUMO is moved from the whole OBO part to the central benzene unit between OBO and the phenyl bridge, while the HOMO is localized on indolocarbazole unit and the phenyl bridge as well. Thus, the overlap between donor and acceptor were increased, resulting in a larger ΔE_{ST} (0.57 eV) than **ICz-*t*BuOBO (41)** (0.56 eV), again showing poor TADF potential (Table 1).

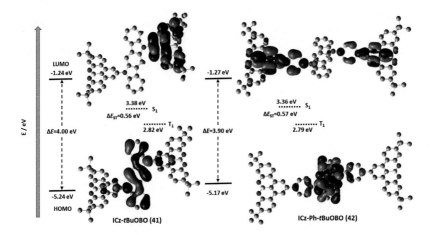

Figure 19. DFT calculations of **ICz-*t*BuOBO (41)** and **ICz-Ph-*t*BuOBO (42)**. Results provided by Dr. Shiv Kumar.

Next, 10*H*-phenoxazine (PXZ), which possesses a central six-membered N-heterocycle, was evaluated as donor for **PXZ-*t*BuOBO (44)** in silico. It shows that the HOMO and LUMO of **PXZ-*t*BuOBO (44)** are dispersed on the PXZ and OBO units, respectively, and the overlap of them is slight. The ΔE_{ST} of **PXZ-*t*BuOBO (44)** is therefore also drastically reduced from 0.65 eV (**Cz-**

*t*BuOBO, 43) to 0.08 eV. Furthermore, when omitting the *tert*-butyl groups, the ΔE_{ST} is further decreased to 0.05 eV (**PXZ-OBO, 45**). The varied ΔE_{ST} could be ascribed to the different dihedral angles between donor and acceptor. When the donor is carbazole or indolocarbazole, the donor to acceptor torsion angles are between 48.41° and 52.28° (Table 1). Using the six-membered N-donor PXZ, the respective angle is increased to 75.48° for **PXZ-OBO (45)** leading to a decreased overlap between donor and acceptor (Figure 20 and Table 1).

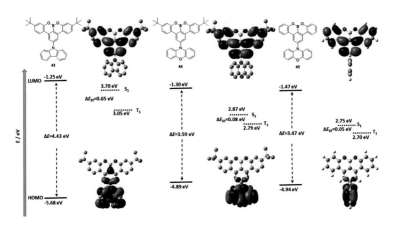

Figure 20. DFT calculations of **Cz-*t*BuOBO (43)**, **PXZ-*t*BuOBO (44)** and **PXZ-OBO (45)**. Results provided by Dr. Shiv Kumar.

Table 1. DFT Calculations of designed molecules **41–45** based on varied N-heterocyclic donors.

Compounds	HOMO/LUMO [eV]	ΔE [eV]	f	S_1 [eV]	T_1 [eV]	ΔE_{ST} [eV]	Dihedral Angles[a] [°]
ICz-*t*BuOBO (41)	−5.24/−1.24	4.00	0.0054	3.38	2.82	0.56	48.41
ICz-Ph-*t*BuOBO (42)	−5.17/−1.27	3.90	0.6507	3.36	2.79	0.57	52.28
Cz-*t*BuOBO (43)	−5.68/−1.25	4.43	0.0001	3.70	3.05	0.65	49.98
PXZ-*t*BuOBO (44)	−4.89/−1.30	3.59	0.0005	2.87	2.79	0.08	-[b]
PXZ-OBO (45)	−4.94/−1.47	3.47	0.0002	2.75	2.70	0.05	75.48[c]

[a] Estimated from optimized structure. [b] Not available. [c] From crystal structure.

These results demonstrated that using the relatively big six-membered rings to increase the dihedral angles between donor and acceptor is an effective method to generate a smaller ΔE_{ST}.[119] Therefore, 10*H*-phenoxazine (PXZ) and another six-membered N-heterocycle, 9,9-dimethyl-9,10-dihydroacridine (DMAC), are further explored as donors in this part.[120] Six emitters were designed and synthesized *via* varying the donors and changing their linking position and count of donor groups (Figure 21).

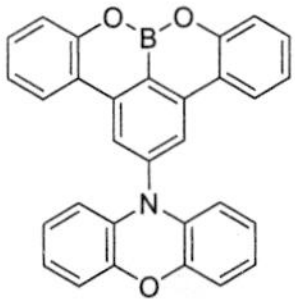

45
PXZ-OBO

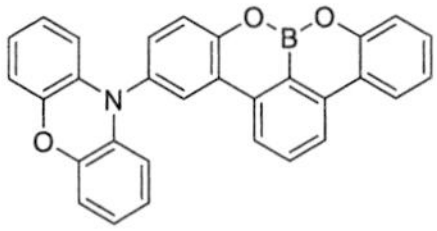

46
5PXZ-OBO

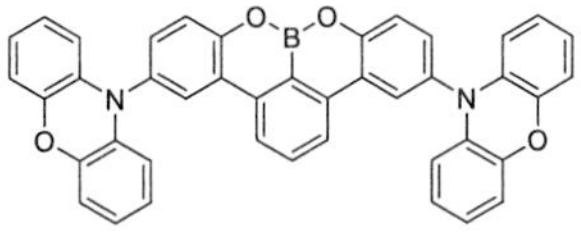

47
DPXZ-OBO

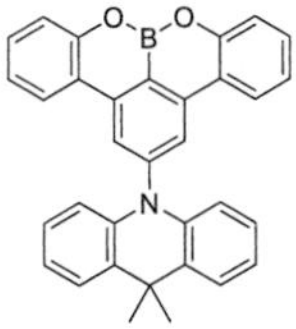

48
DMAC-OBO

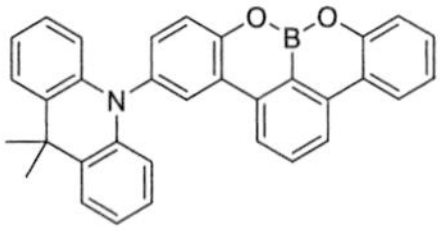

49
5DMAC-OBO

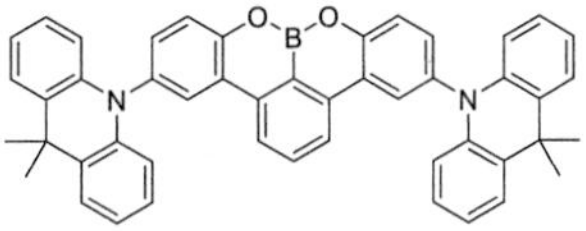

50
DDMAC-OBO

Figure 21. Chemical structures of **PXZ-OBO (45)**, **5PXZ-OBO (46)**, **DPXZ-OBO (47)**, **DMAC-OBO (48)**, **5DMAC-OBO (49)** and **DDMAC-OBO (50)**.

Figure 22 and Table 2 describe the excited state energies and electron density distributions of the HOMOs and LUMOs of the series of emitters. The S_1 and T_1 excited states are calculated from the optimized ground state structure. The ground state geometries match the results of the single crystal data with nearly orthogonal twist between the donor and acceptor units. In addition, the spatial distributions of the HOMOs and LUMOs for all compounds show minimal spatial overlap promoted by the very twisted donor-acceptor orientations induced by the steric repulsion of the hydrogen atoms. The HOMOs disperse on the N-donor, while the LUMOs are localized on the electron-withdrawing OBO units. The well separated frontier orbitals result in small calculated ΔE_{ST} values of 0.04 eV to 0.05 eV, coupled with high optical gaps of 3.34 eV to 3.96 eV, suggesting the potential of PXZ-based molecules as blue TADF emitters. For DMAC-based compounds, their ΔE_{ST} are slightly larger than the corresponding PXZ-based molecules with values from 0.10 eV to 0.18 eV due to the decreased strength when compared to the PXZ donor. Nevertheless, the ΔE_{ST} values for all emitters are small enough to justify further evaluation as TADF emitters. In addition, it was found that two-donor type emitters (donor-acceptor-donor, D-A-D) show smaller ΔE_{ST} than the corresponding one-donor type (donor-acceptor, D-A) emitters.

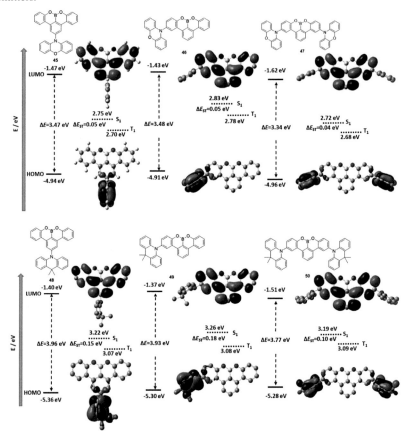

Figure 22. DFT calculations of **PXZ-OBO (45)**, **5PXZ-OBO (46)**, **DPXZ-OBO (47)**, **DMAC-OBO (48)**, **5DMAC-OBO (49)** and **DDMAC-OBO (50)**. Results provided by Dr. Shiv Kumar.

Table 2. DFT Calculations of target molecules **45–50**.

Compounds	HOMO/LUMO [eV]	ΔE [eV]	f	S_1 [eV]	T_1 [eV]	ΔE_{ST} [eV]
PXZ-OBO (45)	−4.94/−1.47	3.47	0.0002	2.75	2.70	0.05
5PXZ-OBO (46)	−4.91/−1.43	3.48	0.0002	2.83	2.78	0.05
DPXZ-OBO (47)	−4.96/−1.62	3.34	0.0003	2.72	2.68	0.04
DMAC-OBO (48)	−5.36/−1.40	3.96	0.0004	3.22	3.07	0.15
5DMAC-OBO (49)	−5.30/−1.37	3.93	0.0003	3.26	3.08	0.18
DDMAC-OBO (50)	−5.28/−1.51	3.77	0.0005	3.19	3.09	0.10

3.1.1.2   Synthesis of OBO-Based Emitters

The six OBO-based derivatives evaluated in silico were synthesized *via* a three-step procedure (Scheme 4–7).

Firstly, 10*H*-phenoxazine (PXZ, **51**) and 9,9-dimethyl-9,10-dihydroacridine (DMAC, **52**) were treated with 1-bromo-3,5-dichlorobenzene under Buchwald-Hartwig cross-coupling conditions to yield **53** and **54**. The remaining two chloro handles were used for a Suzuki cross-coupling process with arylboronic acid to generate dimethoxyteraryl intermediate **55** and **56** (Scheme 4).[121-122]

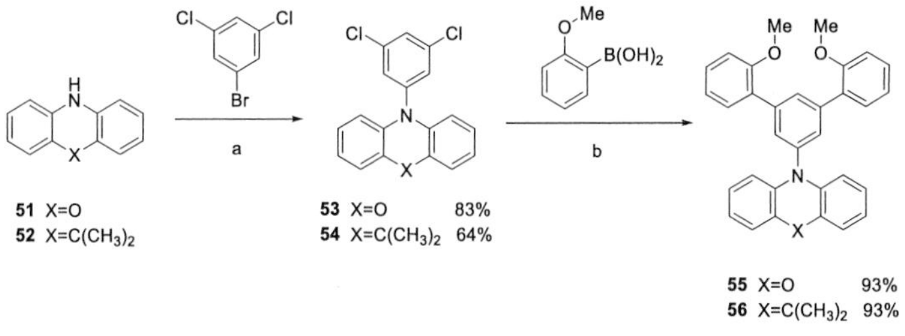

51 X=O
52 X=C(CH₃)₂

53 X=O   83%
54 X=C(CH₃)₂ 64%

55 X=O   93%
56 X=C(CH₃)₂ 93%

Scheme 4. Synthesis of **55** and **56**. a) Pd(OAc)₂, Xantphos, NaO*t*Bu, PhMe, 100 °C, Argon, 12 h. b) Pd(OAc)₂, SPhos, K₃PO₄, THF/H₂O, 80 °C, 12h, Argon.

Through a similar Buchwald-Hartwig cross-coupling process, **57** and **58** were prepared. The following Suzuki cross-coupling for **59** and **60** were conducted between the chloro-contained species and arylboronic ether (Scheme 5).

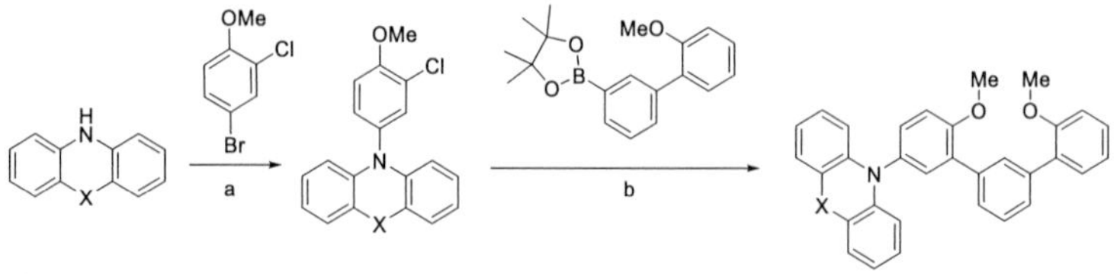

51 X=O
52 X=C(CH_3)$_2$

57 X=O 89%
58 X=C(CH_3)$_2$ 99%

59 X=O 74%
60 X=C(CH_3)$_2$ 92%

Scheme 5. Synthetic route for **59** and **60**. a) $Pd(OAc)_2$, Xantphos, NaO^{t}Bu, PhMe, 100 °C, Argon, 12 h. b) $Pd(OAc)_2$, SPhos, K_3PO_4, PhMe, 100 °C, Argon, 12h.

Based on the methods for **59** and **60**, the dimethoxyteraryl derivatives with two N-donor groups, **61** and **62**, were also obtained (Scheme 6).

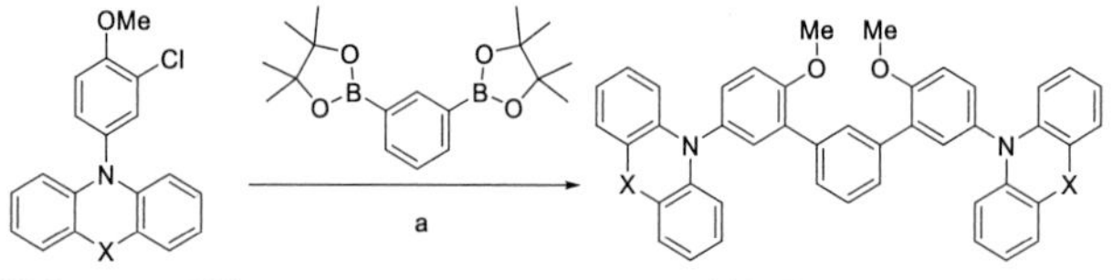

57 X=O 89%
58 X=C(CH_3)$_2$ 99%

61 X=O 66%
62 X=C(CH_3)$_2$ 96%

Scheme 6. Synthesis of **61** and **62**. a) $Pd(OAc)_2$, SPhos, K_3PO_4, PhMe, 100 °C, Argon, 12h.

Lastly, the prepared aryl methyl ethers were used to generate the target OBO-based emitters through a demethylative direct borylation procedure in the presence of BBr_3.[123] The final cyclization proceeded in moderate yields of up to 55–56% for **PXZ-OBO (41)** and **DMAC-OBO (44)** when the donor units were introduced to the central phenyl ring with a direct B–phenyl bond in *para* position. However, the yields dropped to 21–24% when the donor units were in *para* position to the boroxy group (Scheme 7).

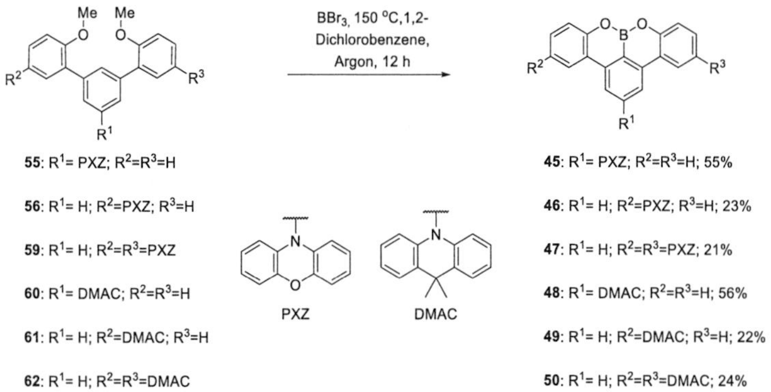

Scheme 7. Synthesis of OBO-fused emitters **PXZ-OBO (45)**, **5PXZ-OBO (46)**, **DPXZ-OBO (47)**, **DMAC-OBO (48)**, **5DMAC-OBO (49)** and **DDMAC-OBO (50)** through a demethylative direct borylation procedure.

The final products were found to be stable under ambient conditions and were easily purified by column chromatography over silica gel. Their structures were fully identified and determined by NMR, MS, HRMS and IR. The absolute configuration of **PXZ-OBO (45)**, **5PXZ-OBO (46)** and **5DMAC-OBO (49)** were also determined by single-crystal X-ray diffraction (Figure 23). The crystal structures reveal that these three molecules possess highly twisted donor-acceptor structures. The dihedral angles between donor and acceptor for the two structural isomers **PXZ-OBO (45)** and **5PXZ-OBO (46)** are tuned remarkably. For **PXZ-OBO (45)**, the angle is 75.48° when PXZ is connected at the *para* position to the boron atom, while it increases to 86.17° for **5PXZ-OBO (46)** where PXZ is connected at the *para* position to the boroxy unit. Further the torsions differ when different donor groups are present. This is shown when comparing **5PXZ-OBO (46)** and **5DMAC-OBO (49)**, which bear the different donor units at the same position on the OBO acceptor with 86.17° (**5PXZ-OBO, 46**) and 80.10° (**5DMAC-OBO, 49**).

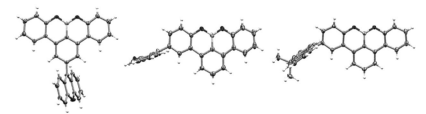

Figure 23. Molecular structure of **PXZ-OBO** (**45**, left), **5PXZ-OBO** (**46**, central) and **5DMAC-OBO** (**49**, right) drawn at 50% probability level.

3.1.1.3   Electrochemical and Photophysical Properties

The electrochemical and photophysical properties of these emitters were performed *via* collaboration with the Professor Eli Zysman-Colman group at the University of St Andrews, UK. All electrochemical and photophysical measurements in this chapter were investigated by Dr. Shiv Kumar.

The electrochemical properties of these six emitters were examined by cyclic voltammetry (CV) and the results are listed in Figure 24 and Table 3. The reductions were assigned to the OBO unit due to its electron-withdrawing property, and the oxidation waves of all compounds were found to be irreversible, which demonstrates that phenoxazine and 9,10 dimethyldihydroacridine radical cations are electrochemically unstable and can subsequently undergo dimerization. In addition, DMAC-based emitters show larger oxidation potentials than the corresponding PXZ-based emitters, which reveals that the phenoxazine is easier to be oxidized.

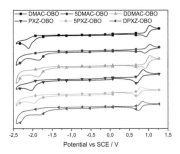

Figure 24. Cyclic Voltammograms of **PXZ-OBO (45)**, **5PXZ-OBO (46)**, **DPXZ-OBO (47)**, **DMAC-OBO (48)**, **5DMAC-OBO (49)** and **DDMAC-OBO (50)** in degassed DCM (scan rate 100 mV s^{-1}).

The UV-Vis absorption and photoluminescence (PL) behaviour of these six OBO-based emitters were investigated in CHCl$_3$ (Figure 25 and Table 3). The OBO-based emitters exhibit similar absorptions and unstructured emission profiles with the maximum absorption bands at 333 nm to 335 nm, which could be ascribed to the intramolecular charge transfer (ICT) absorption process from the PXZ or DMAC donor to the OBO acceptor. The photoluminescence of PXZ-based compounds showed the emission with λ_{PL} of 497 nm to 501 nm, while for DMAC-based compounds a deep blue emission with λ_{PL} of 448 nm to 461 nm was observed.

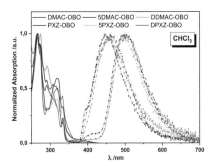

Figure 25. UV-Vis and PL spectra of the boron-containing emitters **PXZ-OBO (45)**, **5PXZ-OBO (46)**, **DPXZ-OBO (47)**, **DMAC-OBO (48)**, **5DMAC-OBO (49)** and **DDMAC-OBO (50)** in CHCl$_3$.

Next, their UV-Vis absorption and PL behaviour in solid state were also investigated (Figure 26 and Table 3). In 10 wt.% doped PMMA films, these six OBO-based derivatives exhibit similar absorptions as in solution, while the emission is hypsochromically shifted to blue and deep blue emission with λ_{PL} of 410 nm to 491 nm, which is desired for TADF emitters. As expected, it was found that the emission could be effectively adjusted by using different donor groups. PXZ-based derivatives show sky-blue emission with λ_{PL} of 471 nm to 491 nm while the emission of DMAC-based derivatives are blue-shifted to λ_{PL} of 410 nm to 446 nm, which is in accordance to the computationally predicted larger band gap of the DMAC group compared to the PXZ group. At the same time, the photoluminescence emission could also be adjusted by different linking type of donor and acceptor. When the donor is introduced to the central phenyl ring of the OBO unit, the emission of **PXZ-OBO (45)** is 491 nm, while it is 481 nm for **5PXZ-OBO (46)** with the donor on the *para* position of the boroxy group. The corresponding DMAC derivatives demonstrate a similar trend. Unfortunately, the PLQYs of all three DMAC-derivatives are ≤10%. This is attributed to the larger band gap of DMAC emitters, which is prone to cause nonradiative decay more easily, compromising efficiency.

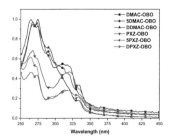 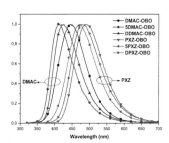

Figure 26. UV-Vis (left) and PL spectra (right) of **PXZ-OBO (45)**, **5PXZ-OBO (46)**, **DPXZ-OBO (47)**, **DMAC-OBO (48)**, **5DMAC-OBO (49)** and **DDMAC-OBO (50)** in 10 wt.% doped films in PMMA.

In contrast in the set of PXZ-derivatives, emitters **PXZ-OBO (45)** and **DPXZ-OBO (47)** show PLQYs of ≥20% justifying further investigation of their TADF properties, with **PXZ-OBO** reaching 46%. Time-resolved PL spectra showed that **PXZ-OBO (45)** and **DPXZ-OBO (47)** possess delayed lifetime components (τ_d) of 850 μs and 1.5 ms, respectively. The long lifetime provides evidence of a delayed fluorescence process of these emitters. With the long delayed lifetime of 850 μs, the PLQY of **PXZ-OBO (45)** reaches up to 46% with λ_{PL} of 491 nm.

Table 3. Electrochemical and photophysical properties of OBO-based compounds

Compounds	HOMO/LUMO[a] [eV]	ΔE^a [eV]	$\lambda_{abs}{}^b/\lambda_{PL}{}^b$ [nm]	$\lambda_{abs}{}^c/\lambda_{PL}{}^c$ [nm]	FWHM [eV][c]	CIE y^c	PLQY [%][c]	τ^c
PXZ-OBO (45)	−5.50/−2.68	2.82	335/500	336/491	0.49	0.36	46	850 μs
5PXZ-OBO (46)	−5.48/−2.71	2.77	333/497	333/471	0.52	0.24	19	N/A*
DPXZ-OBO (47)	−5.52/−2.92	2.60	333/501	334/481	0.50	0.31	20	1.5 ms
DMAC-OBO (48)	−5.71/−2.61	3.10	335/460	337/446	0.56	0.13	8	N/A*
5DMAC-OBO (49)	−5.70/−2.88	2.92	333/448	334/410	0.53	0.10	8	N/A*
DDMAC-OBO (50)	−5.71/−2.85	2.86	334/461	334/420	0.55	0.11	10	N/A*

[a] in CH_2Cl_2 with 0.1 M [nBu$_4$N]PF$_6$ as the supporting electrolyte and Fc/Fc$^+$ as the internal reference. The HOMO and LUMO energies were calculated using the relation $E_{HOMO/LUMO} = - (E^{ox}_{pa,1}/ E^{red}_{pc,1} - E_{Fc/Fc+} - 4.8)eV$, where E^{ox}_{pa} and E^{red}_{pc} are anodic and cathodic peak potentials respectively. $\Delta E^a = - (E_{HOMO}{}^a - E_{LUMO}{}^a)$. [b] in CHCl$_3$ at 298 K. [c] in 10 wt% doped films in PMMA at 298 K. N/A*- not applicable.

In conclusion, six molecules employing a new OBO-fused benzo[fg]tetracene core as acceptor unit and N-heterocycles as donors have been designed and synthesized. They exhibit very good stability and promising blue and deep blue emission with λ_{PL} of 410 nm to 491 nm. Theoretical calculations and experimental studies demonstrate that the photophysical properties of these emitters can be adjusted by utilizing different donors and changing the substituent position and relative number of donor groups. In doped films, the PLQY of these blue emitters can reach up to 46% with delayed lifetime of 850 μs, which shows their potential to be used as TADF emitters for OLEDs.

3.1.2 Triarylboron-Based TADF Emitters

Though OBO-based molecules (see Chapter 3.1.1) show potential to be used as blue and deep blue TADF emitters, the low quantum yields and rather long delayed lifetimes limit their applications for OLEDs.

In 2017, Lee and co-workers designed and synthesized a novel TADF molecule **CzoB (63)** using carbazole as electron-donating unit and triarylboron as electron-accepting group and connecting them *via* the spatial crowding *ortho*-linking method (Figure 27).[124] Triarylboron unit was chosen as acceptor because it is well-known for its strong electron accepting properties as well as for its steric bulkiness particularly when mesityl groups are used.[112] Due to the *ortho*-linking type and bulky nature of the triarylboron, an efficient TADF process was realized in this highly twisted molecule. Furthermore, outstanding blue TADF OLEDs were obtained with an EQE of 24.1%, CIE coordinates of (0.139, 0.198) when employing **CzoB (63)** as emitter. When two *tert*-butyl groups were introduced to the carbazolyl unit yielding emitter **BuCzoB (64)**, the PLQY could be further increased to 91% by inhibiting the concentration quenching.[125] Lu *et al.* subsequently reported the tolyl derivative **B-oTC (65)**.[126] The crystal structure of the emitter indicated strong intramolecular interactions between the *ortho* electron-donating carbazole group and the parallelly oriented mesityl of the triarylboron acceptor unit with a short distance of 2.76–3.55 Å. By combining the charge transfer channels *via* aryl linker and through-space, a very promising TADF emitter was obtained with a small computed ΔE_{ST} of 0.06 eV accompanied by a sufficiently large transition dipole moment. In addition, intermolecular π-π stacking could be inhibited by the rigid structure and bulky substitutions of the aliphatic pending units, leading to reduced concentration quenching. Thus, a very high PLQY of 94% in the neat film and an excellent blue non-doped OLED with an EQE of 19.1% was accessed using **B-oTC (65)** as emitter.

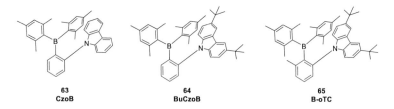

Figure 27. Chemical structures of **CzoB (63)**, **BuCzoB (64)** and **B-oTC (65)**.

In 2015, Lee and co-workers reported that introducing more donors to the donor-acceptor TADF molecular design concept effectively improves the TADF properties (Figure 28).[93] Through the introduction of an additional carbazole-based donor to the parent compound **DCzTrz (66)**, yielding emitter **DCzmCzTrz (67)**, the PLQY in solution was increased from 43% to 84%, and the delayed lifetime was shortened from 31 µs to 9.7 µs. This improve is caused by the accelerated transition rate from triplet to singlet, resulting in more effective TADF emission.

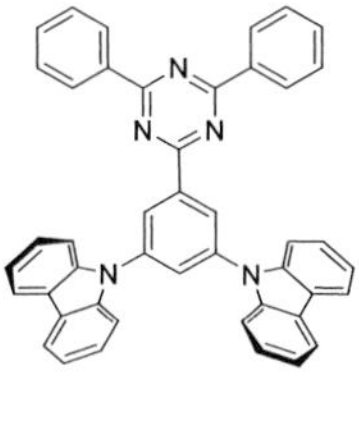

66
DCzTrz

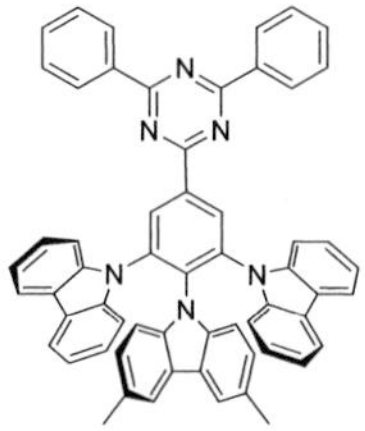

67
DCzmCzTrz

Figure 28. Chemical structures of **DCzTrz (66)** and **DCzmCzTrz (67)**.

3.1.2.1   Molecular Design and DFT Calculation

For this project two triarylboron-based molecules were designed, synthesized and characterized. To combine the observations made in previous literature, the triarylboron unit was linked with a donor at the *ortho* position to enable through-space charge transfer and an increase torsion between the donor and acceptor groups with respect to the bridging phenyl ring, and a second donor was added to the *para* position to increase the RISC rates. In detail, a carbazole unit was introduced to the *para* position, and another *tert*-butyl-substituted carbazole group to the *ortho* position of the boron atom on the central phenyl ring to construct emitter **CzCzoB (68)**. For **ICzDCzoB (69)**, an indolocarbazole group was used as a bridge to connect two triarylboron units in an acceptor-donor-acceptor type arrangement (Figure 29).

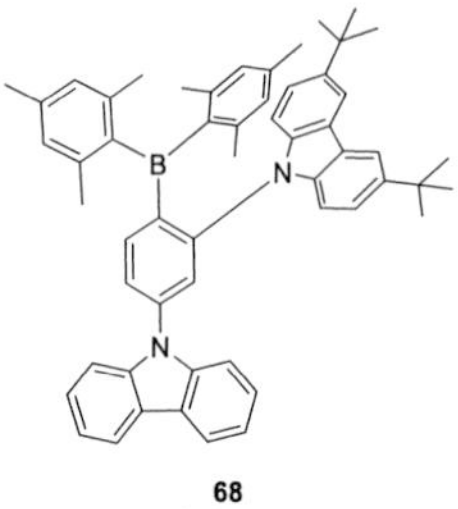

Figure 29. Chemical structures of **CzCzoB** (**68**) and **ICzDCzoB** (**69**).

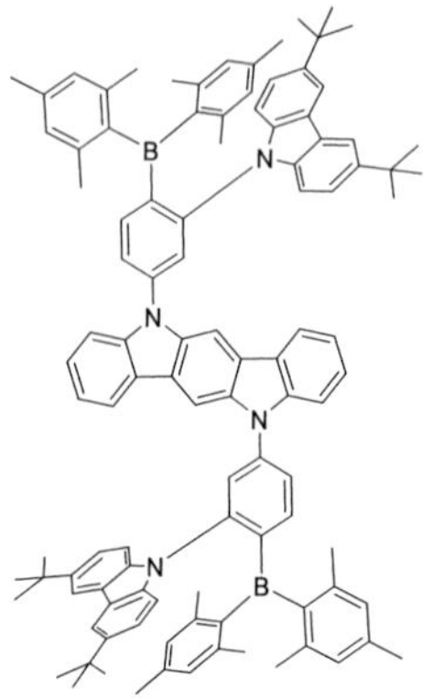

DFT calculations were performed to assess the electronic structure of these two molecules. The frontier orbitals (HOMO/LUMO) and energy levels are shown in Figure 30 and Table 4. The HOMO of **CzCzoB** (**68**) is located on the appended carbazole unit at the *ortho* position to the acceptor, while the LUMO is dispersed on the boron-based acceptor unit itself. This indicates that a charge-transfer transition is dominant between donor and acceptor units. For compound **ICzDCzoB** (**69**), the HOMO is localized on the central indolocarbazole unit. The different HOMO distributions of **CzCzoB** (**68**) and **ICzDCzoB** (**69**) show that the electron-donating strength of indolocarbazole is stronger than of carbazole. As expected, the LUMO is localized on the boron-based units for both molecules, which represents a charge-transfer transition between donor and acceptor units.

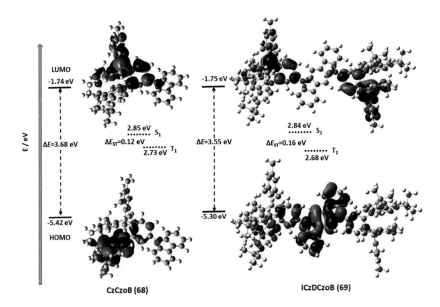

Figure 30. HOMO and LUMO orbitals of **CzCzoB** (**68**, left) and **ICzDCzoB** (**69**, right).

Table 4. DFT Calculations of **CzCzoB** (**68**) and **ICzDCzoB** (**69**).

Compounds	HOMO/LUMO [eV]	ΔE [eV]	f	S_1 [eV]	T_1 [eV]	ΔE_{ST} [eV]
CzCzoB (68)	−5.42/−1.74	3.68	0.015	2.85	2.73	0.12
ICzDCzoB (69)	−5.30/−1.75	3.55	0.044	2.84	2.68	0.16

3.1.2.2 Synthesis of Triarylboron-Based Emitters

The two emitters designed and evaluated in silico were successfully synthesized by a three-step procedure (Scheme 8–11).

In the first step, a nucleophilic aromatic substitution was conducted on 1,4-dibromo-2-fluorobenzene (**70**) with 3,6-di-tert-butyl-9*H*-carbazole (**71**) as the nucleophile to generate the compound **72** with NaH as base in dry DMF at 100 °C (Scheme 8). The two *tert*-butyl groups were introduced to the carbazolyl unit to increase solubility and inhibit the concentration quenching of the final emitters.

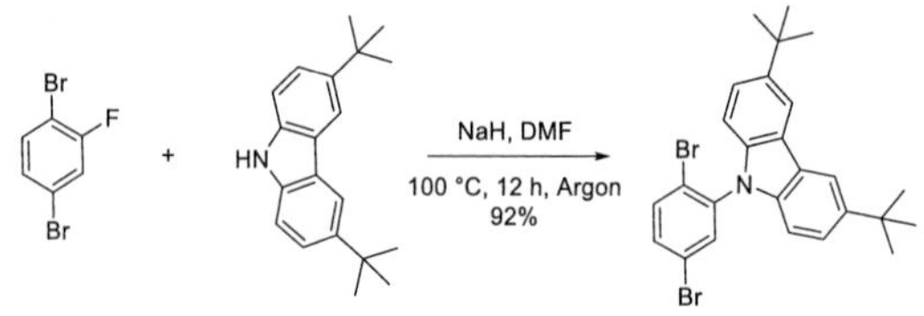

Scheme 8. Synthesis of carbazolyl derivative 72.

Then, the key intermediate 74 was obtained by selective *ortho* lithiation of 3,6-di-tert-butyl-9-(2,5-dibromophenyl)-9-*N*-carbazole (72), followed by quenching with dimesitylboron fluoride (Mes$_2$BF, 73).[127] Owed to the electron-donating and therefore a *n*-butyllithium coordinating effect induced by carbazole-based unit, the *ortho*-bromine is lithiated preferentially (Scheme 9). The remaining bromine atom acts as the final functional group for the last step.

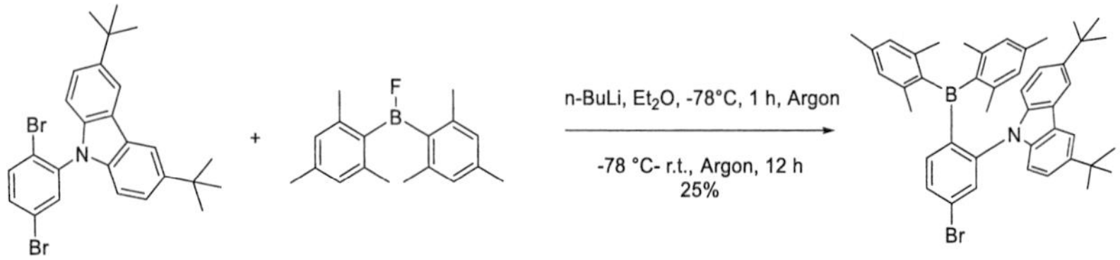

Scheme 9. Synthesis of triarylboron compound 74.

Lastly, the emitter CzCzoB (68) was obtained through a Pd-catalyzed Buchwald-Hartwig cross-coupling reaction between triarylboron-based intermediate 74 and carbazole (75) with a yield of 85% (Scheme 10).

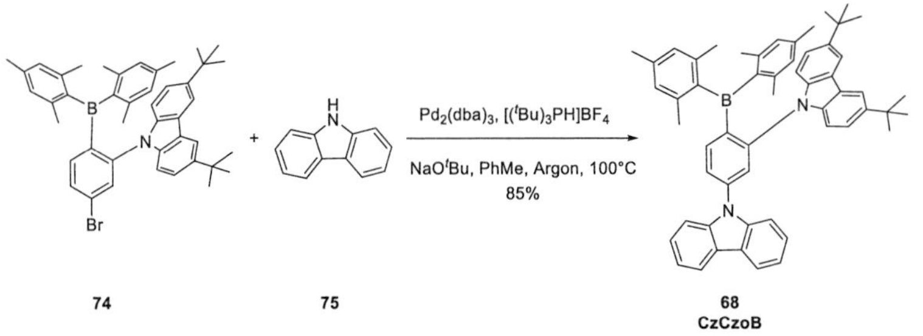

Scheme 10. Synthesis of the emitter **CzCzoB (68)**.

Under identical conditions **ICzDCzoB (69)** was prepared in 62% yield when indolocarbazole **76** was used as cross-coupling partner for two equivalences of boron-based intermediate (Scheme 11).

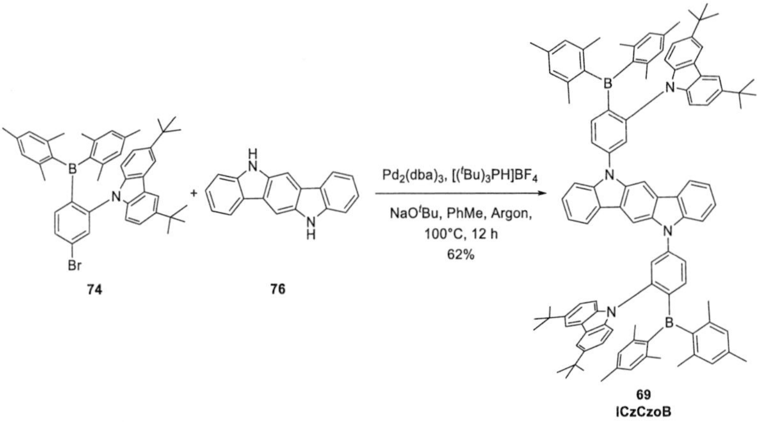

Scheme 11. Synthesis of the emitter **ICzDCzoB (69)**.

3.1.2.3 Photophysical Properties

The photophysical properties of these two emitters were performed *via* collaboration with Professor Anna Köhler at the University of Bayreuth, Germany. All photophysical measurements in this chapter were investigated by Eimantas Duda.

Figure 31 presents the absorption (UV-Vis) spectra of **CzCzoB (68)** and **ICzDCzoB (69)** and the results are listed in Table 5. Peaks between 290 to 365 nm (346 nm, 365 nm) for **CzCzoB (68)** and between 290 to 340 nm (323 nm, 338 nm) for **ICzDCzoB (69)** can be ascribed to local transitions in the donor and acceptor moieties. The low intensity peak for **CzCzoB (68)** between 400 and 440 nm can be assigned to the charge-transfer absorption between donor and acceptor units. The peak at around 403 nm for **ICzDCzoB (69)** arises from the n-π* transition of the indocarbazole moiety.[117, 128]

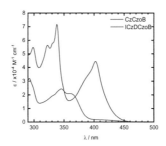

Figure 31. UV-Vis spectra of **CzCzoB (68)** and **ICzDCzoB (69)** in toluene measured at room temperature.

In Figure 32, the photoluminescence emission is presented for molecule **CzCzoB (68)** both in solution and film in steady-state regime and also at a specific delay and gating time. The maximum of the steady-state emission at room temperature is 489 nm in solution and 493 nm in 10 wt.% DPEPO film. Emission with a delayed time of 50 ms and gating time of 50 ms at 77 K is assigned to phosphorescence. The maximum of the phosphorescence spectrum is at 480 nm in solution and 501 nm in film. The redshift of emission in film both for steady-state emission and phosphorescence is likely to be due to a higher molecular planarity in the solid state and, as a consequence, higher degree of charge delocalization across the molecule. The steady-state emission at room temperature does not have structure in solution and film while phosphorescence shows only a very badly pronounced structure. Steady-state emission at room temperature is likely to be a charge-transfer emission. The HOMO and LUMO orbital distributions in Figure 30 for **CzCzoB (68)** indeed show the charge-transfer transition for the singlet state, where the HOMO is localized on the appended carbazole part of the molecule while the LUMO is localized on the boron-based unit of the molecule.

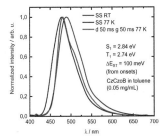 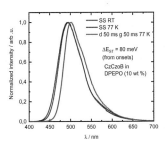

Figure 32. Emission spectra of **CzCzoB (68)** in toluene (left) and DPEPO film (right) at RT and 77 K taken at a steady-state (SS) condition or at a specific delay 50 ms and gating 50 ms. The concentrations for solutions and films are 0.05 mg/mL and 10 wt %, respectively. λ_{exc} = 340 nm.

For compound **ICzDCzoB (69)**, the maximum of the steady-state emission is at 485 nm in toluene and 486 nm in film at room temperature (Figure 33). There is no significant shift in emission wavelength going from solution to film. The steady-state emission is similar to that of **CzCzoB (68)** indicating a similar origin for the steady-state emission at room temperature. Based on the absence of structure, the origin of this emission is likely to be a charge-transfer transition between the appended carbazole and boron-based unit. This is supported by the distribution of the HOMO and LUMO orbitals shown in Figure 30. Unlike in compound **CzCzoB (68)**, where phosphorescence does not appear to have any pronounced structure, phosphorescence has a very clear structure for compound **ICzDCzoB (69)** with a position of the highest energy peak at 476 nm. In film, the position of phosphorescence is redshifted and the structure becomes less pronounced. Based on the position and structure, the origin of phosphorescence is very different for **ICzDCzoB (69)** compared to **CzCzoB (68)**. A structured nature of the spectrum is an indication for a locally-excited emission. This can be supported in the publication by Ting *et al.*, where the molecules investigated have the same indolocarbazole unit as in **ICzDCzoB (69)**, however, different groups are attached to nitrogen atoms in the indolocarbazole unit. The phosphorescence spectrum reported has the position of the highest energy peak at 475 nm as well as a vibrational structure, which is identical to the phosphorescence of **ICzDCzoB (69)**. Therefore, the origin of phosphorescence of **ICzDCzoB (69)** is the locally-excited emission from the indolocarbazole unit in the molecule.

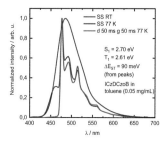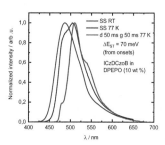

Figure 33. Emission spectra of **ICzDCzoB (69)** in toluene (left) and DPEPO film (right) at RT and 77 K taken at a steady-state (SS) condition or at a specific delay 50 ms and gating 50 ms. The concentrations for solutions and films are 0.05 mg/mL and 10 wt %, respectively. λ_{exc} = 340 nm.

Next, the photoluminescence quantum yields of the two emitter were measured in solution and in film. The increased PLQYs were observed when the measurements were conducted under nitrogen atmosphere instead of air, which is a common character of TADF emitters. Especially, the efficiency of **CzCzoB (68)** reaches up to 93%, higher than in solution (54%), leading to a very promising emitter for OLEDs. In comparison with **CzCzoB (68)**, the PLQY of **ICzDCzoB (69)** is not as high as **CzCzoB (68)**, but it can still reach up to 67%. In addition, the emission of **ICzDCzoB (69)** is at 486 nm, which is bluer than **CzCzoB (68)** (493 nm).

In order to further understand the emission processes, the singlet and triplet states of these two emitters were investigated and the results were summarized in Table 5. The singlet and triplet states of **CzCzoB (68)** in solution are at 2.84 eV and 2.74 eV as determined from the onsets of the spectra, thus the ΔE_{ST} between them is 0.10 eV. When doped in host material, the ΔE_{ST} is further reduced to 0.08 eV, which is beneficial to enable the TADF process. For **ICzDCzoB (69)**, the ΔE_{ST} is 0.09 eV in solution as determined from the peaks of the spectra and 0.07 eV in film as determined from the onsets of the spectra. The small ΔE_{ST} values of **CzCzoB (68)** and **ICzDCzoB (69)** are consistent to those obtained from DFT calculation, which are 0.12 eV for **CzCzoB (68)** and 0.16 eV for **ICzDCzoB (69)** (Table 4).

Time-resolved photoluminescence of the two emitters in film was performed to reveal and confirm their TADF properties. The results are summarized in Table 5. As shown in Figure 34, two emission regimes can be identified, namely, prompt emission and delayed emission. The lifetime of prompt emission is 28 ns for **CzCzoB (68)** and 18 ns for **ICzDCzoB (69)**. For both compounds the emission in the microsecond range is present and increases with increasing temperature, which clearly shows the TADF nature of these two emitters. The lifetime of this emission is 3.1 µs for **CzCzoB (68)** and 4.0 µs for **ICzDCzoB (69)** when fitted using a stretched exponential with β = 0.4. The ratio of delayed and prompt emission is 0.9 for **CzCzoB (68)** and 0.6 for **ICzDCzoB (69)** at 300 K. Compared to a similar emitter reported by Lee and co-workers, **BuCzoB (64)**,[125] the delayed lifetime of **CzCzoB (68)** was drastically shortened from 23.0 µs (**BuCzoB, 64**) to 3.1 µs, showing to be more promising for efficient OLEDs.

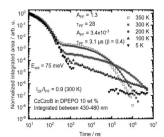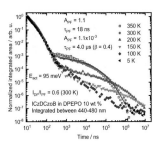

Figure 34. CzCzoB (68) and **ICzDCzoB (69)** doped at 10 wt% into DPEPO. Left: The decay of the luminescence signal of **CzCzoB (68)**, spectrally integrated from 430–480 nm, taken at different temperatures as indicated. Right: The decay of the luminescence signal of **ICzDCzoB (69)**, spectrally integrated from 440–480 nm, taken at different temperatures as indicated. λ_{exc} = 355 nm.

Combining the temperature activated delayed emission and the small ΔE_{ST} as well as activation energies which are consistent with the S-T gap, the origin of the delayed emission is assigned to TADF.

Table 5. Photophysical properties of **CzCzoB (68)** and **ICzDCzoB (69)**.

Compounds	$\lambda_{abs}{}^c$ [nm]	$\lambda_{PL}{}^d$ [nm]	PLQY[%]^e	$S_1/T_1{}^g$	$\Delta E_{ST}{}^i$	$\tau_p// \tau_d$
CzCzoB (68)^a	346, 365	489/478/480	54(32)	2.84/2.74	0.10	32/-
CzCzoB (68)^b	-	493/494/501	93(83)	2.74/2.66	0.09	28/3.1^k
ICzDCzoB(69)^a	323, 338, 403	485/478/476	35(26)	2.70^h/2.61^h	0.08	20/-
ICzDCzoB (69)^b	-	486/504/509	67(51)	2.70/2.63	0.07	18/4.0^k

^{a} in toluene. ^{b} in 10 wt% DPEPO film. ^{c} UV-Vis in toluene. ^{d} photoluminescence measured in the condition of steady-state (SS) at room temperature, steady-state (SS) at 77 K and delayed 50 ms and gating 50ms at 77 K, repecially. ^{e} PL efficiency measurements taken under N_2 atmosphere and in air. λexc = 315 nm. ^{fg} Obtained from the onset of the PL steady-state and delayed emission spectra at 77 K, λ_{exc} = 340 nm. ^{h} Obtained from the maximum of the PL spectra at 77 K. ^{i} $\Delta E_{ST} = E(S_1)$ $- E(T_1)$. ^{j} λ_{exc} = 355 nm at 300 K under vacuum, prompt component τ_p using a single exponential decay. ^{k} λ_{exc} = 355 nm at 300 K under vacuum, delayed component τ_d obtained using a stretched exponential y = A*exp(-(t/τ_d)$^\beta$) + y$_0$, where β = 0.4.

In summary, two triarylboron-based emitters were investigated. Through the modification of a TADF emitter with the introduction of a carbazole unit, the PLQY of **CzCzoB (68)** reached up to 93% with a very short delayed lifetime of 3.1 μs, which is one of the shortest lifetime of triarylboron-derived TADF emitters reported so far. For the other emitter bearing an indolocarbazole as central donor **ICzDCzoB, 69**), sky-blue emission with a PLQY of 67% and a delayed lifetime of 4.0 μs was realized. The excellent photophysical properties of these two emitters suggest that they are good candidates for efficient TADF OLEDs and the further electroluminescent studies are in preparation.

3.2 Indolocarbazole-Based Donors for TADF Emitters

As shown in last chapter 5,11-dihydroindolo[3,2-b]carbazole (**ICz, 76**) is a particularly interesting donor unit for the construction of TADF molecules, because it possesses a larger planar rigid conjugation structure and a stronger electron-donating strength than a carbazolyl unit. Actually, **ICz (76)** is just one isomer of the indolocarbazole family. Depending on the shared bond position and orientation of the fused indole unit to the carbazole, five indolocarbazole isomers can be constructed, namely 5,11-dihydroindolo[3,2-*b*]carbazole (**ICz, 76**), 11,12-dihydroindolo[2,3-*a*]carbazole (**77**), 5,7-dihydroindolo[2,3-*b*]carbazole (**78**), 5,8-dihydroindolo[2,3-*c*]carbazole (**79**), and 5,12-dihydroindolo[3,2-*a*]carbazole (**80**).[129-130] Most of them and their respective derivatives have been investigated and show promising applications in biology and materials science.

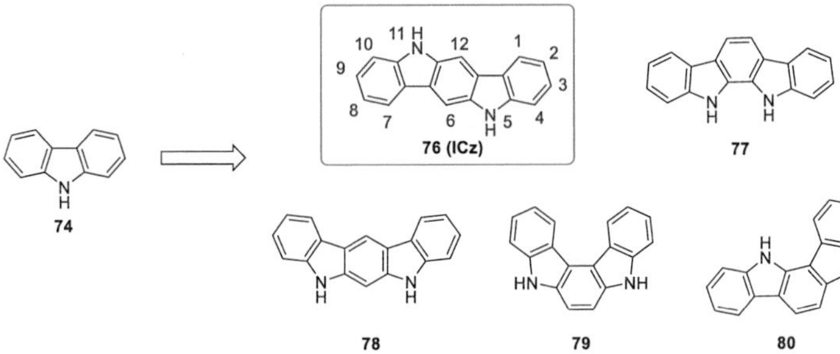

Figure 35. The isomers of indolocarbazole and nomenclature of **ICz (76)**.

3.2.1 Indolocarbazole-Based TADF emitters

In the TADF OLEDs field, Adachi and co-workers reported the first purely organic TADF emitter **PIC-TRZ (5)** with a small ΔE_{ST} of 0.08 eV by using a derivative of the mirror symmetric indolocarbazole **78** as donor in 2011.[54] Employing **PIC-TRZ (5)** as emitter (Figure 36), a green OLED was established with emission at 506 nm and EQE of 5.3%. Furthermore, indolocarbazole **80** was subsequently used to construct emitter **PIC-TRZ2 (81)**, which possesses a further decreased ΔE_{ST} of 0.003 eV due to the well separated HOMO and LUMO caused by the increased steric twist.[131] In 2018, the group of Wang also reported a green OLED with emission at 496 nm and a very high EQE of 30.0% by employing a TADF emitter **IndCzpTr-2 (35)** based on the point symmetric indolocarbazole (ICz, 76). Additionally, a blue TADF OLED with emission of 472 nm was realized with **IndCzpTr-1 (82)** as emitter. Unfortunately, the EQE of the device was only 14.5%.[110] In the same year, Duan and co-workers investigated two TADF molecules derived from the two indolocarbazole isomers **78** and **80**. In that series, **32aICTRZ (83)**, showed excellent TADF properties with a high internal quantum efficiency of 93% and a maximum EQE of 25.1%.[132] The TADF emitter **23bICTRZ (84)** based on **78** showed sky-blue emission, but the EQE dropped to 9.8%. Moreover, TADF molecule **32aICTRZ (83)** could also be used as a host material for green TADF-OLEDs with low roll-off device performances (EQE of 26.2% at 5000 cd m^{-2}).

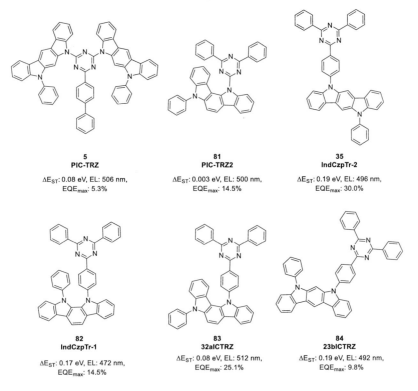

5
PIC-TRZ
ΔE_{ST}: 0.08 eV, EL: 506 nm,
EQE$_{max}$: 5.3%

81
PIC-TRZ2
ΔE_{ST}: 0.003 eV, EL: 500 nm,
EQE$_{max}$: 14.5%

35
IndCzpTr-2
ΔE_{ST}: 0.19 eV, EL: 496 nm,
EQE$_{max}$: 30.0%

82
IndCzpTr-1
ΔE_{ST}: 0.17 eV, EL: 472 nm,
EQE$_{max}$: 14.5%

83
32aICTRZ
ΔE_{ST}: 0.08 eV, EL: 512 nm,
EQE$_{max}$: 25.1%

84
23bICTRZ
ΔE_{ST}: 0.19 eV, EL: 492 nm,
EQE$_{max}$: 9.8%

Figure 36. Chemical structures and performances of known indolocarbazole-based TADF emitters, **PIC-TRZ (5)**, **PIC-TRZ2 (81)**, **IndCzpTr-2 (35)**, **IndCzpTr-1 (82)**, **32aICTRZ (83)** and **23bICTRZ (84)**.

As shown above, it has been proven that indolocarbazoles are promising donors for excellent green TADF emitters. However, blue emission with good device performances has not been studied fully, particularly not in the indolocarbazole series. It is noteworthy that all so far reported emitters in the literature adhere to the principle of combining one or more donor units to one acceptor unit. The unique feature of the indolocarbazoles lays in the possibility to invert this preference to a mono-donor, di-acceptor combination. Therefore, this part is focused on indolocarbazole-based emitters for blue TADF emission and the evaluation of a mono-donor, di-acceptor combination for TADF emitter design.

3.2.1.1 Molecular Design and DFT Calculation

First, three new acceptor-donor-acceptor (A-D-A) linear molecules were designed using the point symmetrical indolocarbazole **ICz (76)** as donor, which allows the orientation of the acceptor groups in a linear 180° orientation, and two symmetric triazine-based derivatives as acceptors (Figure 37). The *tert*-butyl groups were introduced to di-phenyltriazine to tune the electron-withdrawing strength, extend the length of the target emitters and maintain solubility.[133] The choice for **ICz (76)** as the central donor is due to its large planar rigid conjugation structure, suitable electron-donating strength and the two antisymmetric reactive positions, which can facilitate the construction of linear molecules for the horizontal orientation preference of the emitter in solid films.[107, 117] In addition, restricting the HOMO on the central indolocarbazole moiety is advantageous to maintain a large energy gap between HOMO and LUMO, needed to obtain blue emission. For the parent emitter **ICzTRZ (85)**, the two phenyl rings between **ICz (76)** and triazine were used as bridges without additional electronically or sterically active groups, to show the magnitude of the HOMO/LUMO interaction.[118] Furthermore, for the emitter **FICzTRZ (86)**, fluorine atoms were added to the phenyl bridges at the *ortho* position to the indolocarbazole donor unit to tune the dihedral angles between donor and acceptor as reported by Jiang *et al.*[134] In addition, the **O***t***BuICzTRZ (87)** derivative of **ICzTRZ (85)** with more *tert*-butyl groups was constructed to further vary the electron-accepting strength and increase the solubility.

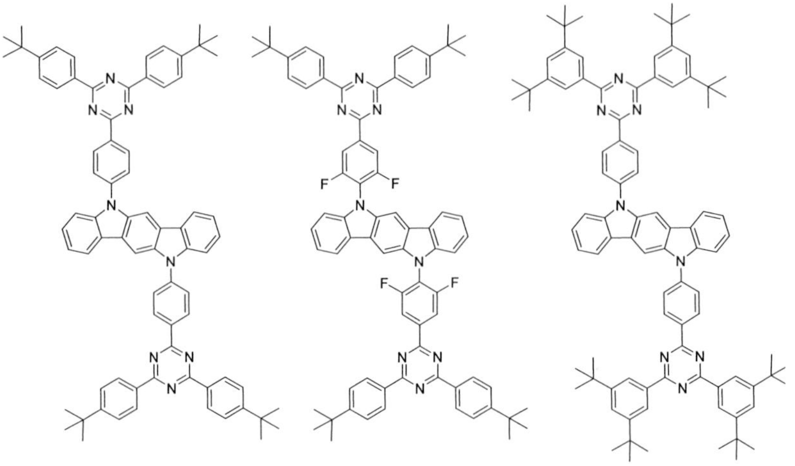

85
ICzTRZ

86
FICzTRZ

87
OtBuICzTRZ

Figure 37. Chemical structures of the target molecules based on indolocarbazole **ICz (76)**, **ICzTRZ (85)**, **FICzTRZ (86)** and **OtBuICzTRZ (87)**.

Density functional theory (DFT) calculations were performed to establish whether these molecules could be promising TADF emitters by employing the Gaussian 09 revision D.018 suite. Firstly, the PBE0 functional with the standard Pople 6-31G (d,p) basis set was used to optimize the geometries of the molecular structures in the ground state in gas phase. Next, time-dependent DFT calculations were performed with the Tamm–Dancoff approximation (TDA) based on the optimized molecular structures. The GaussView 5.0 software was employed to visualize the molecular orbitals and provide initial insights on the nature of the modified properties induced by different donors. The frontier orbitals (HOMO/LUMO) and energy levels of these molecules are shown in Figure 38, Figure 39 and Table 6.

The HOMOs of these three molecules are mainly dispersed on the indolocarbazole donor unit, while the LUMOs are localized on the two triazine parts and the phenyl bridges (Figure 38 and 39). Moreover, HOMOs and LUMOs are partly overlap on the phenyl bridge. Emitter **ICzTRZ (85)** presents a theoretical ΔE_{ST} of 0.22 eV combined with a high oscillator strength (*f*) of 0.72 a.u. (Table 6).

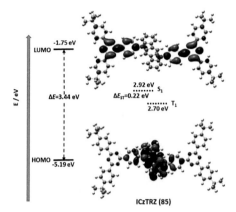

Figure 38. DFT calculations of **ICzTRZ (85)**. Results provided by Ettore Crovini.

As shown in Figure 39, the derivatives of the parent emitter **ICzTRZ (85)** show varying trends. The ΔE_{ST} of **FICzTRZ (86)** is decreased to 0.16 eV from 0.22 eV due to the addition of fluorine atoms on the linking bridges. The ΔE_{ST} of **OtBuICzTRZ (87)** is slightly decreased to 0.21 eV. It is also potential to realize TADF emission.

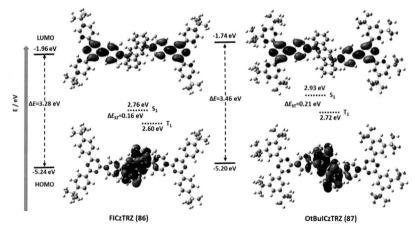

Figure 39. DFT calculations of **FICzTRZ (86**, left) and **OtBuICzTRZ (87**, right). Results provided by Ettore Crovini.

Table 6. DFT calculations of **ICz**-based emitters **85–87**.

Compounds	HOMO/LUMO [eV]	ΔE [eV]	f	S_1 [eV]	T_1 [eV]	ΔE_{ST} [eV]
ICzTRZ (85)	−5.19/−1.75	3.44	0.7212	2.92	2.70	0.22
FICzTRZ (86)	−5.24/−1.96	3.28	0.5114	2.76	2.60	0.16
OtBuICzTRZ (87)	−5.20/−1.74	3.46	0.7200	2.93	2.72	0.21

3.2.1.2 Synthesis of Indolocarbazole-Based Emitters

The target indolocarbazole-based molecules were synthesized by a four-step convergent synthetic procedure (Scheme 12–14). In the first step, the indolocarbazole donor unit **ICz** (**76**) was synthesized *via* an acid promoted cyclization using commercially available 3,3-diindolylmethane (**88**) as precursor[135-136] in multigram scale (10.0 g) (Scheme 12). The TADF donor **76** was isolated in 64% yield after 1 h.

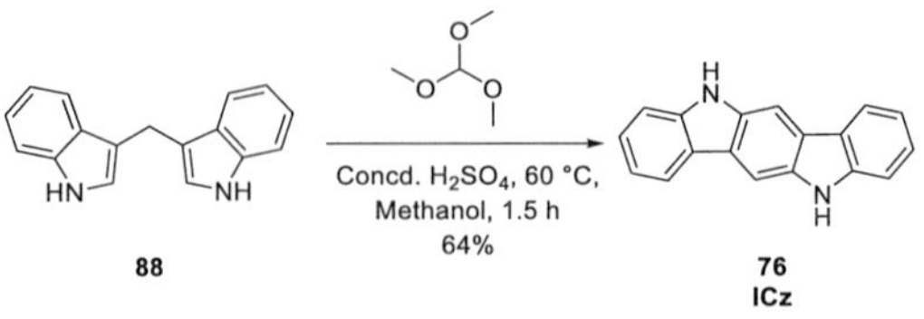

Scheme 12. Synthesis of the isomer of indolocarbazole **ICz** (**76**).

Secondly, the triazine-based acceptor parts **91** and **92** were obtained through Grignard reaction between two equivalents of the respective magnesiated aryl bromides **89** and **90** with 2,4,6-trichloro-1,3,5-triazine in moderate yields.[133] Next, the acceptor was extended with a 4-fluorophenyl or 3,4,5-fluorophenyl linker *via* Suzuki cross-coupling of the respective 4-fluorophenyl boronic acids to generate the final fluoro-acceptor units **93**, **94** and **95** in good to excellent yields (Scheme 13).

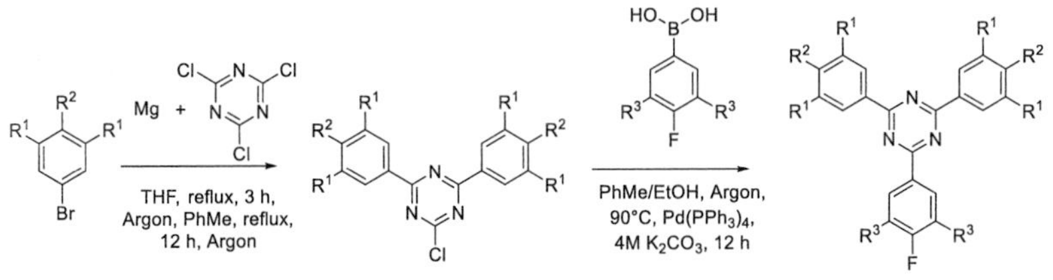

HO‚B‚OH

THF, reflux, 3 h,
Argon, PhMe, reflux,
12 h, Argon

PhMe/EtOH, Argon,
90°C, Pd(PPh₃)₄,
4M K₂CO₃, 12 h

89: R¹=H, R²=ᵗBu
90: R¹=ᵗBu, R²=H

91: R¹=H, R²=ᵗBu; 58%
92: R¹=ᵗBu, R²=H; 63%

93: R¹=H, R²=ᵗBu, R³=H; 87%
94: R¹=ᵗBu, R²=H, R³=H; 91%
95: R¹=H, R²=ᵗBu, R³=F; 90%

Scheme 13. Synthesis of the *tert*-butyl group decorated triazine-based acceptors, **92**, **93** and **94**.

Finally, the **ICz (76)** donor and the respective acceptor units were connected *via* nucleophilic aromatic substitution in a 1:2 ratio with tripotassium phosphate as base in DMSO at 120°C to yield **ICzTRZ (85)**, **FICzTRZ (86)** and **OtBuICzTRZ (87)** in moderate yields (Scheme 14).

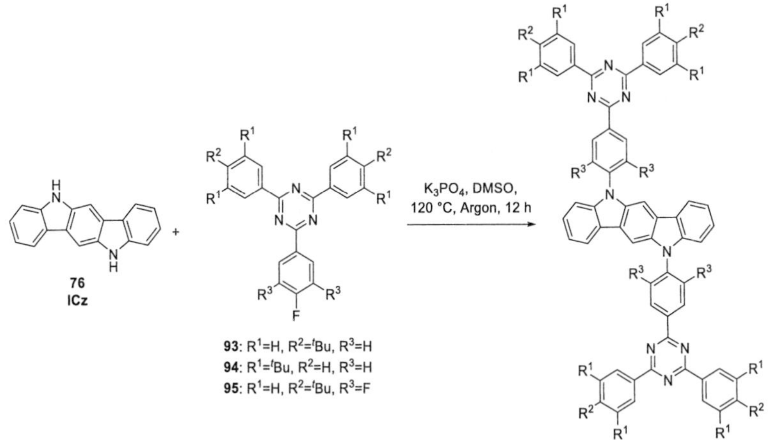

76
ICz

K₃PO₄, DMSO,
120 °C, Argon, 12 h

93: R¹=H, R²=ᵗBu, R³=H
94: R¹=ᵗBu, R²=H, R³=H
95: R¹=H, R²=ᵗBu, R³=F

85: R¹=H, R²=ᵗBu, R³=H; 66%
87: R¹=ᵗBu, R²=H, R³=H; 68%
86: R¹=H, R²=ᵗBu, R³=F; 57%

Scheme 14. Synthesis of the emitters **ICzTRZ (85)**, **FICzTRZ (86)** and **OtBuICzTRZ (87)** through the nucleophilic aromatic substitution.

These three emitters were fully characterized by NMR, mass spectra (MS and HRMS), IR. The absolute configuration of **ICzTRZ (85)** was further determined by single crystal analysis. As shown in Figure 40, the structure of this emitter possesses a twisted donor-acceptor structure. The angle between the phenyl bridge and the indolocarbazole donor unit is 45.81°, while the angle between the phenyl bridge and triazine core of the acceptor is only 9.17°.

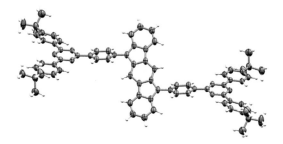

Figure 40. Molecular structure of the linear emitter **ICzTRZ (85)** drawn at 50% probability level.

3.2.1.3 Photophysical and Electrochemical Properties

Photophysical properties of ICzTRZ (85)

The photophysical and electrochemical properties of these emitters were performed *via* collaboration with the Professor Eli Zysman-Colman group at the University of St Andrews, UK. All photophysical and electrochemical measurements in this chapter were investigated by Ettore Crovini.

First of all, host material screenings were conducted to find the optimal host to realize the best photoluminescent performances. The PPT, DPEPO, PMMA and the carbazole-derived mCP and mCBP were tested because of their excellent charge transport abilities and high energy states. The doping concentration, which is also very crucial to the emission yield, was varied from 1 wt% to 20 wt%.

As shown in Table 7, increased PLQYs were observed when the measurements were conducted under nitrogen atmosphere instead of air condition, which is a common characteristic of TADF emitters. The PLQYs drastically depend on the host material and concentration. For instance, in mCP the quantum yield was increased from 51.6% (doped at 10 wt%) to 68.0% (doped at 5 wt%). By changing host materials, the quantum yield could be further increased from 40.0% (doped at 5 wt%) in PPT to 70.0% (doped at 5 wt%) in mCBP.

Table 7. Host optimization of spin-coated films for **ICzTRZ (85)**.

Entry	Host material	Doping concentration	$\Phi_{PL}^{thin\ film}$ air/ %	$\Phi_{PL}^{thin\ film}$ nitrogen/ %
1		1 wt%	60.6	67.0
2		2 wt%	59.2	65.0
3	mCP	3 wt%	53.2	57.4
4		4 wt%	63.0	68.0
5		5 wt%	50.9	53.5
6		10 wt%	48.8	51.6
7		3 wt%	16.0	24.0
8	PPT	5 wt%	34.0	40.0
9		10 wt%	25.0	30.0
10		15 wt%	28.0	32.0
11		3 wt%	18.0	22.0
12	DPEPO	10 wt%	19.0	23.8
13		20 wt%	32.0	35.0
14	PMMA	10 wt%	28.0	31.0
15	mCBP	5 wt%	58.8	70.0

Figure 41 and Table 8 present the absorption (UV-Vis) spectra of **ICzTRZ (85)** in neat thin film and doped in host materials. The peaks from 300 to 340 nm could be ascribed to local transitions in the donor and acceptor moieties. The peaks at around 400 nm (390–406 nm) are n-π* transitions of the indocarbazole moiety as reported by Shi *et al.*[117] The photoluminescence emission maximum is at 489 nm in neat film and it shifts to 470 nm, 472 nm and 479 nm when the emitter is doped in other unpolar hosts such as PMMA, mCP and mCBP, respectively. Based on the host screening results, the best photoluminescent performances were obtained in mCBP with a PLQY of 58.8% under air and 70.0% under N_2 atmosphere.

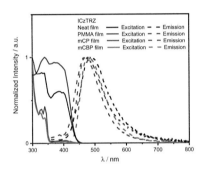

Figure 41. UV-Vis absorption and PL spectra of **ICzTRZ (85)** in neat thin film and doped in host materials.

Table 8. Photophysical properties of **ICzTRZ (85)** in neat thin film and doped in host materials.

Entry	host	λ_{abs} in DCM / nm	λ_{PL} thin film / nm	Φ_{PL} thin film air / %	Φ_{PL} thin film nitrogen / %	τ_p / ns	τ_d / μs
1	neat	340, 395	489	30.0	32.0	8.67	No delayed
2	PMMA (10 wt%)	340, 390	470	28.0	31.0	11.34	252.83
3	mCP (3 wt%)	330, 340, 405	472	53.2	57.4	4.80	113.60
4	mCBP (5 wt%)	327, 340, 406	479	58.8	70.0	9.01	121.09

Furthermore, transient PL measurements showed prompt fluorescence with a triexponential decay kinetics and a fitted lifetime of 9.01 ns (τ_1=1.52 ns (3.34%), τ_2=6.52 ns (60.97%), τ_3=13.97 ns (35.69%)) followed by a long-delayed component with biexponential decay kinetics with a lifetime of 121.09 μs (τ_1=19.13 μs (26.05%), τ_2=157.01 μs (73.95%)) (Figure 42). In addition, the delayed emission was affected by the environment of the films. As shown in Figure 42, the prompt component in neat film (8.67 ns) and in mCBP (9.01 ns) is similar, but the delayed component was not observed in neat film.

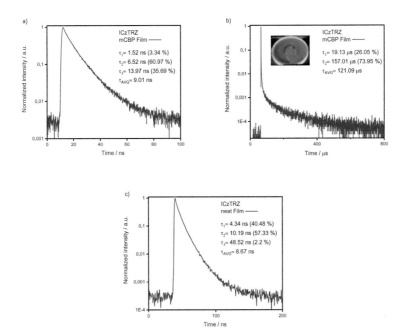

Figure 42. a) Prompt fluorescence decay of **ICzTRZ (85)** in spin-coated film with 5 wt% of the emitter in mCBP (λ_{exc}= 340 nm). b) Delayed fluorescence decay of **ICzTRZ (85)** in spin-coated film with 5 wt% of the emitter in mCBP (λ_{exc}= 340 nm). c) Prompt fluorescence decay of **ICzTRZ (85)** in neat film (λ_{exc}= 340 nm).

The TADF nature was confirmed by measuring the temperature dependence of the delayed component which shows an increased intensity with higher temperature due to the activation of the RISC process by the higher thermal energy given to the system (Figure 43). The ΔE_{ST} was obtained from the onset of the prompt fluorescence and delayed phosphorescence spectra, measured from an evaporated film of 5 wt% **ICzTRZ (85)** in mCBP at 77 K, with a value of 0.23 eV, which is in line with the value obtained from TDA-DFT calculations (0.22 eV).

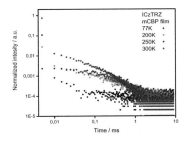 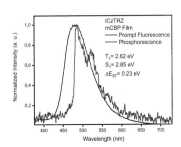

Figure 43. Left) Temperature dependence of the delayed component (λ_{exc}= 340 nm). Right) Prompt fluorescence spectra (RT), phosphorescence spectra (77K), T_1 and S_1 energy level measured with Prompt camera (λ_{exc}= 340 nm).

Electrochemical Properties of ICzTRZ (85)

The electrochemical properties of **ICzTRZ (85)** were examined by cyclic voltammograms (CV) and differential pulse voltammetry (DPV) in degassed DCM (Figure 44), and the results are listed in Table 9. The reversible oxidation peak at +0.96 eV and an irreversible reduction peak at −1.83 eV (with regards to SCE) were observed. The reduction progress was assigned to the triazine acceptor part due to its electron-withdrawing property. The oxidation wave was irreversible, which suggests that indolocarbazole radical cations are electrochemically unstable and can subsequently undergo dimerization similarly to carbazole. The FMO values were calculated from the onset of the oxidation and reduction peaks, respectively, giving values of −5.68 eV for the HOMO and −3.17 eV for the LUMO (Table 9).

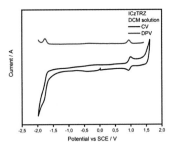

Figure 44. Cyclic Voltammetry (CV) and Differential Pulse Voltammetry (DPV) of **ICzTRZ (85)** corrected against the Saturated Calomel Electrode (SCE).

Table 9. Electrochemical properties of **ICzTRZ (85)**.

Compound	Oxidation potential [eV]	HOMO [eV][a]	Reduction potential [eV]	LUMO [eV][a]	ΔE [eV]
ICzTRZ (85)	0.96	−5.68	−1.83	−3.17	2.51

[a] HOMO and LUMO were obtained from the onset of the redox potentials in cyclic voltammetry.

Photophysical Properties of FICzTRZ (86)

In contrast to **ICzTRZ (85)**, **FICzTRZ (86)** exhibits similar absorption (UV-Vis) in neat thin films and doped in host materials as shown in Figure 45 and Table 10. Peaks in the range of 300–340 nm can be ascribed to local transitions in the donor and acceptor moieties while the peaks at around 400 nm is a n-π* transition of the indocarbazole moiety. However, due to the introduction of fluorine atoms, the emission maximum of **FICzTRZ (86)** red-shifts to 523 nm from 489 nm (**ICzTRZ, 85**). The emission could also be tuned based on the host from 523 nm (in neat film) to 522 nm (in PMMA) and 491 (in mCP). Though the PLQYs exhibit the trend of increase from air to inert atmosphere in neat film and doped in host materials, the highest PLQY is only 39.5% under nitrogen in 5 wt% mCP. In addition, the ratio of delayed component is only 1% in 5 wt% mCP, which suggests that emitter **FICzTRZ (86)** is not a promising TADF emitter.

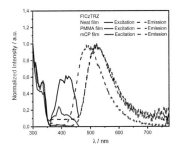 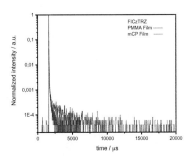

Figure 45. UV-Vis absorption and PL spectra (left) and photoluminescence decay curves (right) of **FICzTRZ (86)**.

Table 10. Photophysical properties of **FICzTRZ (86)** in neat thin film and doped in host materials.

Entry	host	λ_{abs} in DCM /nm	λ_{PL} thin film / nm	Φ_{PL} thin film air / %	Φ_{PL} thin film nitrogen / %	τ_p / ns	τ_d / μs
1	neat	340, 405	523	27.8	30.6	10.1	No delayed
2	PMMA (10 wt%)	340, 395	522	21.4	26.5	9.5	0.15(1%)
3	mCP (5 wt%)	340, 390	491	36.8	39.5	9.2	1(1%)

Photophysical Properties of OtBuICzTRZ (87)

For **OtBuICzTRZ (87)**, which contains eight *tert*-butyl groups, the emission maximum shifts from 474 nm (in neat film) to 483 nm (in PMMA) and 479 nm (in mCP) (Figure 46). In comparison to **ICzTRZ (85)**, the emission maxima stay very constant in the sky-blue region. The PLQYs are significantly increased from 10.7 % in neat film to 65.9% in mCP film. As expected, the prompt and delayed emission in mCBP were both observed with the lifetimes of 5.8 ns and 190.8 μs, respectively. As mentioned before, the delayed lifetime of 171.9 μs is in the range of typical TADF molecules, confirming the TADF properties of this emitter (Table 11).

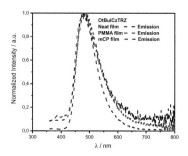 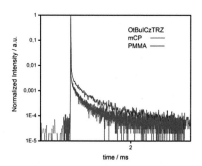

Figure 46. PL spectra (left) and photoluminescence decay curves (right) of **OtBuICzTRZ (87)**.

Table 11. Photophysical properties of **OtBuICzTRZ (87)** in neat thin films and doped in host materials.

Entry	host	$\lambda_{PL}^{thin\ film}$ / nm	$\Phi_{PL}^{thin\ film}$ air / %	$\Phi_{PL}^{thin\ film}$ nitrogen / %	τ_p / ns	τ_d / μs
1	neat	474	9.7	10.7	4.1	No delayed
2	PMMA (10 wt%)	483	27.0	31.1	7.8	438.3
3	mCP (10 wt%)	479	63.0	65.9	5.8	190.8

3.2.1.4 Horizontal Orientation

Due to the excellent TADF properties of **ICzTRZ (85)** and **OtBuICzTRZ(87)**, the anisotropy factor (a) of the emitters was determined in several host materials. Unfortunately, the vapor-deposited films of **OtBuICzTRZ (87)** could not be fabricated because its increased molecular weight is too high to be sublimated. For **ICzTRZ (85)**, doped films with 10 wt% **ICzTRZ (85)** in three different hosts were obtained *via* evaporation. The orientation of the emitter molecule was then measured using polarization- and angle-dependent luminescence spectroscopy, which was then followed by optical simulation. **ICzTRZ (85)** exhibits a nearly-completely horizontal orientation in all the three tested host materials (Figure 45) with the best results obtained from the DPEPO doped-film with a value of preference of 94% (a = 0.06) (Figure 47 and Table 12).

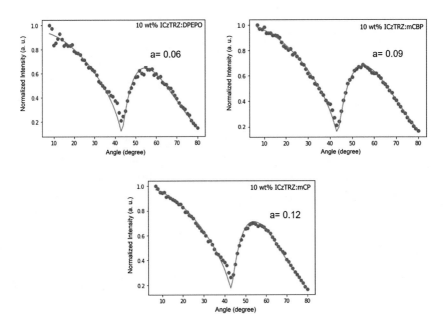

Figure 47. Anisotropy factor of **ICzTRZ (85)** in several host materials obtained by using polarization- and angle-dependent luminescence spectroscopy (the simulation curve in orange).

Table 12. Orientation properties of **ICzTRZ (85)** in host materials.

Host	Doping/wt%	a^a	θ^b	S^c
DPEPO	10	0.06	0.94	1.41
mCBP	10	0.09	0.91	1.365
mCP	10	0.12	0.88	1.32

^{a} anisotropy factor; ^{b} fraction of horizontal dipole (θ= 1-a); ^{c} orientation order parameter (S = [3(1-a)]/2).

In order to further understand the role of the *tert*-butyl groups for the horizontal orientation, a similar molecule without the *tert*-butyl groups, **ICzTRZ-0 (96)** (Figure 48), was designed and synthesized.

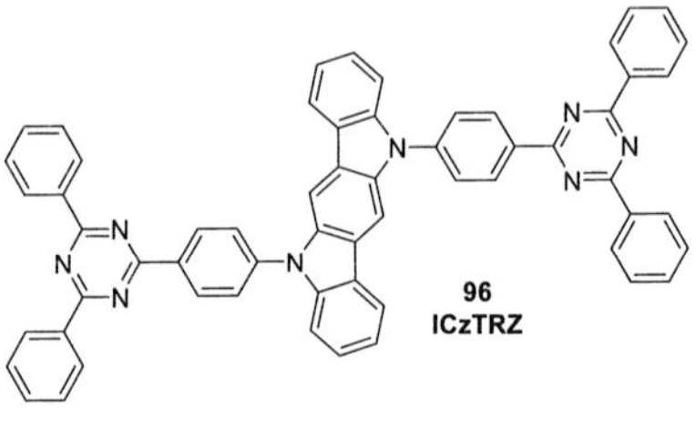

Figure 48. Chemical structure of molecule **ICzTRZ-0 (96)**.

The synthetic route for **ICzTRZ-0 (96)** was based on the previously discussed synthetic path to **ICzTRZ (85)**. A nucleophilic aromatic substitution was conducted between **ICz (76)** and commercially available 2-(4-fluorophenyl)-4,6-diphenyl-1,3,5-triazine (**98**) to yield emitter **ICzTRZ-0 (96)** in 80% yield (Scheme 15).

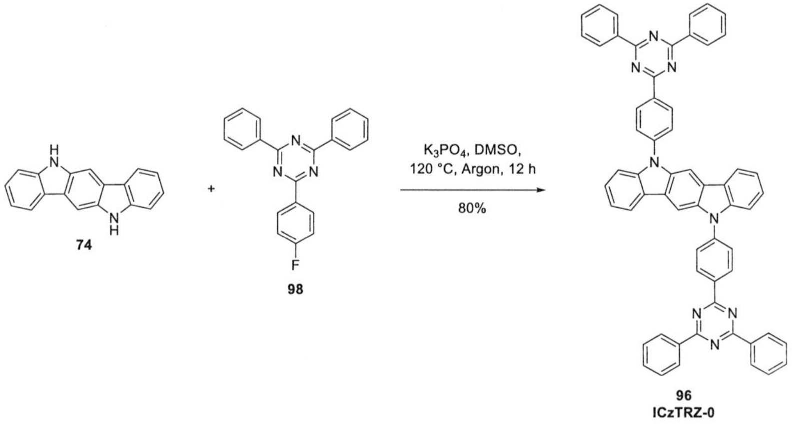

Scheme 15. Synthesis of **ICzTRZ-0 (96)**.

DFT calculations were conducted to reveal the HOMO/LUMO distributions and energy states of **ICzTRZ-0 (96)** (Figure 49 and Table 13). Compared to **ICzTRZ (85)**, the LUMO level of **ICzTRZ-0 (96)** is reduced from −1.75 eV to −1.86 eV, and the band gap between HOMO and

LUMO of **ICzTRZ-0 (96)** is also slightly reduced from 3.44 eV to 3.39 eV. However, ΔE_{ST} remains almost identical. Based on these results, the introduction of *tert*-butyl groups resulted in the slight increase of the LUMO level, accompanied with almost unaffected ΔE_{ST}, which is within the limits of observing blue TADF emission.

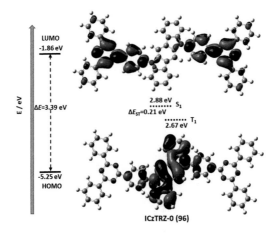

Figure 49. DFT calculations of **ICzTRZ-0 (96)**. Results provided by Ettore Crovini.

Table 13. DFT calculations of **ICzTRZ-0 (96)** and **ICzTRZ (85)**.

Compounds	HOMO/LUMO[a] [eV]	ΔE^a [eV]	f^b	S_1^b [eV]	T_1^b [eV]	ΔE_{ST}^b [eV]
ICzTRZ-0 (96)	−5.25/−1.86	3.39	0.6781	2.88	2.67	0.21
ICzTRZ (85)	−5.19/−1.75	3.44	0.7212	2.92	2.70	0.22

Next, the anisotropy factor (a) of **ICzTRZ-0 (96)** was measured in several host materials (Figure 50 and Table 14). Unlike **ICzTRZ (85)**, which shows excellent horizontal orientation in most host materials (θ as high as 94%), the best horizontal orientation preference of **ICzTRZ-0 (96)** was observed in mCBP with a decreased preference θ of 88%. Based on these results, the introduction of *tert*-butyl groups to the acceptor not only increases the LUMO level, but also enhances the horizontal orientation preference in films, both of which are beneficial for the goal of highly efficient blue TADF OLEDs.

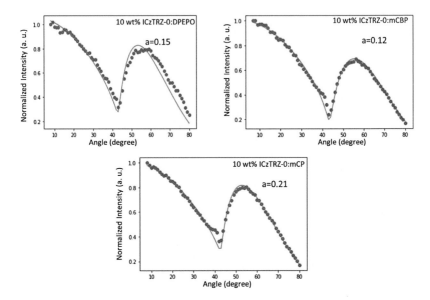

Figure 50. Anisotropy factor of **ICzTRZ-0 (96)** in several host material obtained using polarization- and angle-dependent luminescence spectroscopy.

Table 14. Horizontal orientation properties of **ICzTRZ-0 (96)** and **ICzTRZ (85)** in host materials.

Compounds	Host	Doping/wt%	a^a [eV]	θ^b	S^c
ICzTRZ-0 (96)	DPEPO	10	0.15	0.85	1.275
	mCBP	10	0.12	0.88	1.32
	mCP	10	0.21	0.79	1.185
ICzTRZ (85)	DPEPO	10	0.06	0.94	1.41

[a] anisotropy factor; [b] fraction of horizontal dipole (θ= 1-a); [c] orientation order parameter (S = [3(1-a)]/2).

3.2.1.5   OLED Fabrication

The OLED device performances of **ICzTRZ (85)** were performed *via* collaboration with the Professor Ifor D. W. Samuel group at the University of St. Andrews, UK. All the device measurements in this chapter were investigated by Dr. Paloma L. dos Santos.

To evaluate the potential of **ICzTRZ (85)** in devices, several OLEDs were fabricated. The device presented in Figure 51 is representative of the OLEDs tested. mCBP was chosen as the host material because it gave the best PLQY in doped films while also retaining a high degree of horizontal orientation. The optimized device structure was: ITO/TAPC (40 nm)/TCTA (10 nm)/5 wt% **ICzTRZ (85)**:mCBP (20 nm)/DPEPO (10 nm)/TmPyPb (50 nm)/LiF (0.6 nm)/Al (100 nm), where indium tin oxide (ITO) is the anode, 4,4'-cyclohexylidenebis[N,N-bis(4-methylphenyl)benzenamine] (TAPC) and tris(4-carbazoyl-9-ylphenyl)amine (TCTA) act as hole transport layer, 3,3'-Di(9H-carbazol-9-yl)-1,1'-biphenyl (mCBP) is the host. Bis[2-(diphenyl)benzene (TmPyPB) acts as electron-transporting material, and LiF modifies the work function of the aluminum cathode (Figure 51a). The doping ratio was optimized at 5 wt%.

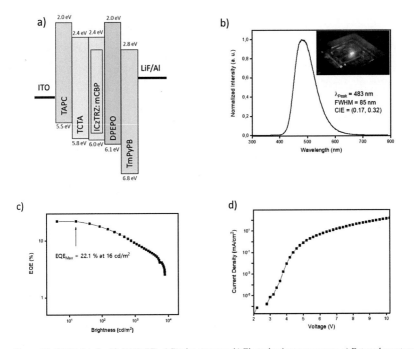

Figure 51. OLED data for **ICzTRZ (85)**. a) Device structure. b) Electroluminescent spectra. c) External quantum efficiencies (EQE) versus brightness. d) Current density versus voltage curves.

The electroluminescence (EL) spectra shows blue emission with a peak maximum at 483 nm and associated CIE chromaticity coordinates of (0.17, 0.32) (Figure 51b). Figure 51c describes EQE versus luminance curves. The OLED shows an EQE_{max} of 22.1% at 16 cd m^{-2}. With this EQE_{max} value, the high PLQY of 70% in mCBP and considering the charge balance as 100% and that essentially 100% of the triplet are being efficiently harvested and converted to singlet excitons *via* the TADF mechanism, the out-coupling efficiency is estimated at around 31%, which is higher than that usually found in TADF OLEDs. Thus, it can be inferred that the emitter is showing some degree of alignment in the device. At 100 cd m^{-2} the EQE_{100} value is only slightly decreased to 17.3%. At 1000 cd m^{-2} the EQE_{1000} values decrease to 9.1%. The device shows a low turn-on voltage of ~3.5 eV and brightness levels reaching 7805 cd/m^2 (at EQE = 2.5%).

Table 14. Device performances of **ICzTRZ (85)**.

Device	$V_{on}{}^{a}$/ V	$EQE_{max}{}^{b}$/ %	$EQE_{100}{}^{bc}$/ %	$EQE_{1000}{}^{bc}$/ %	$Brg_{max}{}^{d}$/ cd/m^2	CIE / (x,y)	$\lambda_{EL}{}^{f}$/ nm
ICzTRZ (85): mCBP	3.5	22.1	17.3	9.1	7805	(0.17, 0.32)	483

a: Von= Turn on voltage; b: EQE= external quantum efficiency; c: EQE_{100} and EQE_{1000} refer to values taken at 100 cd/m^2 and 1000 cd/m^2; d: Brg= brightness; e: CIE= Internationale de L'Éclairage coordinates; f: EL= electroluminescence.

In summary, a set of four TADF emitters based on indolocarbazole was designed, synthesized and characterized. Amongst this series **ICzTRZ (85)**, presented the highest PLQY in doped films and almost complete horizontal orientation in all the hosts that were tested. These properties encouraged the fabrication of OLED devices, which led to a high-performance OLED with blue emission (CIE coordinates of 0.17, 0.32) and an above-average EQE_{max} of 22.1%.

3.2.2 Dimerization Strategy for TADF Emitters

The conjugation extended derivative of carbazole, **ICz (76)**, has exhibited high potential to be used as donor for highly efficient blue TADF emitters **ICzTRZ (85)**. For the carbazole-based structure, another method to enlarge the conjugation is the dimerization through a single bond between two carbazole units, rather than a fused indole at the carbazole. As shown in Figure 52, the dimer **102** or even polymers of carbazole can be accessed *via* an oxidative coupling procedure.[137-139] Dimer **102** was also reported as donor to construct TADF emitters by Adachi and Lee.[140-141]

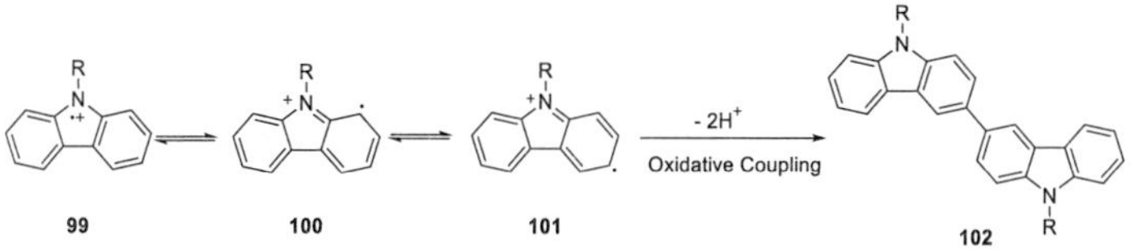

Figure 52. Oxidative coupling procedure to generate the dimer of carbazole-based derivatives.

3.2.2.1 Molecular Design and DFT Calculation

Considering the structural similarity of **ICz (76)** and carbazole, the analogous dimer **DICzTRZ (104)** of the evaluated promising TADF emitter **ICzTRZ (85)** (see Chapter 3.2.1) is explored in this part (Figure 53). For comparison, the dimer **DCzTRZ (103)** based on carbazole unit is also investigated.

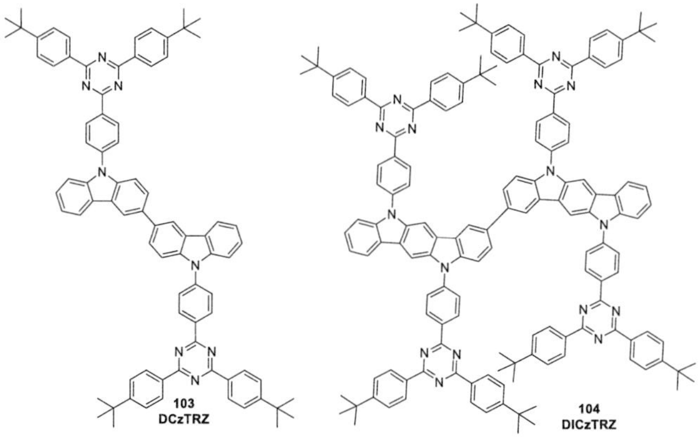

Figure 53. Chemical structure of **DCzTRZ (103)** and the dimer **DICzTRZ (104)**.

DFT calculations were performed to assess the modeling properties of the dimer of carbazole-based molecule **DCzTRZ (103)**. As shown in Figure 54, the LUMO of **DCzTRZ (103)** is localized on the two triazine parts and the phenyl bridges while the HOMO is mainly dispersed on the biscarbazole donor unit. In comparison with **ICzTRZ (85)**, the ΔE_{ST} of **DCzTRZ (103)** is slightly increased to 0.24 eV due to the increased overlap between HOMO and LUMO. Due to the very large structure of **DICzTRZ (104)**, the DFT calculation method we used for **DCzTRZ (103)** was not suitable for it. The further modeling study for **DICzTRZ (104)** is under investigation.

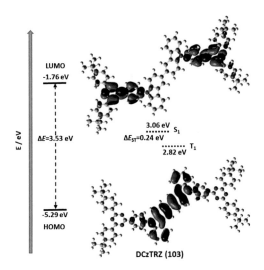

Figure 54. DFT calculations of **DCzTRZ (103)**. Results provided by Ettore Crovini.

Table 15. DFT calculations of **DCzTRZ (103)** and **ICzTRZ (85)**.

Compounds	HOMO/LUMO [eV]	ΔE [eV]	f	S_1 [eV]	T_1 [eV]	ΔE_{ST} [eV]
DCzTRZ (103)	−5.29/−1.76	3.53	0.6781	3.06	2.82	0.24
ICzTRZ (85)[a]	−5.19/−1.75	3.44	0.7212	2.92	2.70	0.22

[a] From Chapter 3.2.1

3.2.2.2 Synthesis of Dimer Emitters

As the carbazole dimers are constructed by oxidative coupling, the monomeric emitters were synthesized first, and then dimerized to the final emitters. In the first step, the monomer of carbazole derivative **106** was synthesized through a palladium-catalyzed Suzuki cross-coupling reaction between (4-(9*H*-carbazol-9-yl)phenyl)boronic acid (**105**) and 2,4-bis(4-(tert-butyl)phenyl)-6-chloro-1,3,5-triazine (**91**) with moderate yield (Scheme 16).

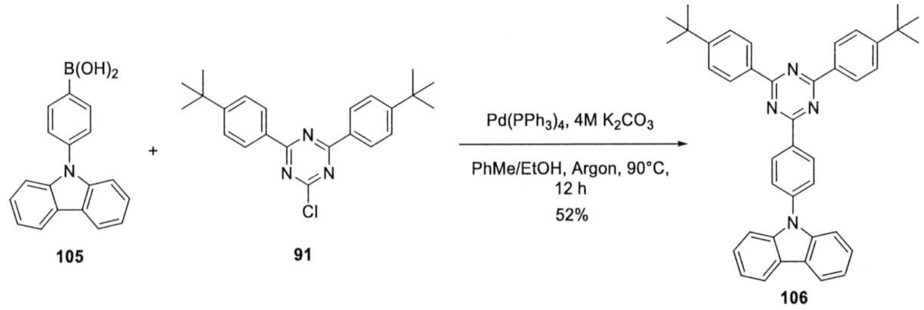

Scheme 16. Synthesis of the monomer of carbazole-based derivative **106**.

Next, the dimer of **DCzTRZ (103)** was obtained with a yield of 58% *via* oxidative coupling using FeCl$_3$ as the catalysis under argon atmosphere at room temperature (Scheme 17).[137]

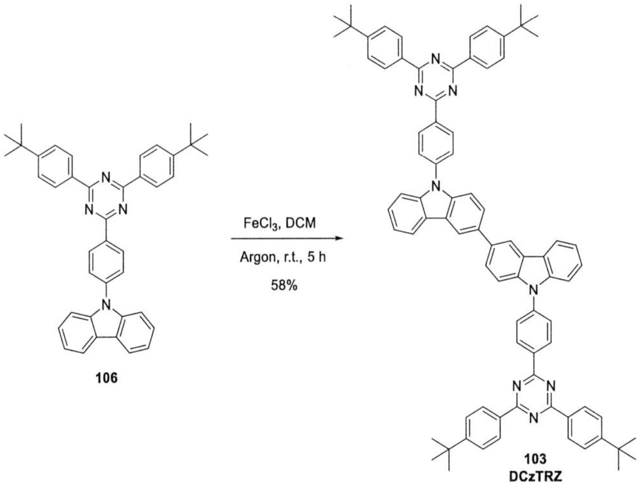

Scheme 17. Synthesis of the dimer **DCzTRZ (103)**.

Then, the identical oxidative coupling conditions were applied to synthesize the dimer **DICzTRZ** (**104**) of the previously synthesized emitter **ICzTRZ** (**85**). Unfortunately, no desired product of dimer **DICzTRZ** (**104**) was observed at room temperature after 12h. Prolonging the reaction time to 48 h did not yield any product. It was found that when the temperature was slightly increased to 40 °C for 12 h, the dimer was formed and could be isolated with a yield of 20%. No full conversion was obtained with considerable amounts of the starting monomer (**85**) remaining. To drive this conversion closer to completion, the temperature was further increased to 60 °C. Surprisingly, after 12 h all monomer was consumed and the dimer **DICzTRZ** (**104**) could be isolated in a yield of 66% (Table 16).

Table 16. Screening conditions towards the dimer **DICzTRZ** (**104**).

Entry	Temperature	Time	Yield
1	r.t.	12 h	No conversion
2	r.t.	48 h	No conversion
3	40 °C	12 h	Dimer (**104**, 20%)
4	60 °C	12 h	Dimer (**104**, 66%)

3.2.2.3   Photophysics of Dimer Emitters

Photoluminescence properties of **DCzTRZ (103)** in neat film and in different host materials were investigated. The photoluminescence emission maximum at 474 nm in neat film was blue-shifted to 470 nm and 460 nm when the emitter was doped in unpolar hosts PMMA and mCP, respectively. Also, in mCP, the quantum yield was increased to 57.4% with pure blue emission at 460 nm under nitrogen. Time-resolved photoluminescence behavior of **DCzTRZ (103)** in films was performed to reveal its emission properties. As shown in Figure 55 and Table 17, the emission decay curve with a multiexponential fitting showed the prompt decay component in the nanosecond regime from 6.37 ns to 10.17 ns. Unfortunately, no delayed component was observed in the emission, showing no TADF characteristics of **DCzTRZ (103)** in these tested host materials.

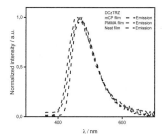 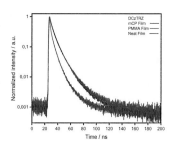

Figure 55. PL spectra (left) and photoluminescence decay curves (right) of **DCzTRZ (103)**.

Table 17. Photophysical properties of **DCzTRZ (103)** in neat thin film and doped in host materials.

Entry	host	$\lambda_{PL}^{\text{thin film}}$ / nm	$\Phi_{PL}^{\text{thin film}}$ air / %	$\Phi_{PL}^{\text{thin film}}$ nitrogen / %	τ_p / ns	τ_d / μs
1	neat	474	31.1	34.1	8.93	No delayed
2	PMMA (10wt%)	470	26.7	33.6	10.17	No delayed
3	mCP (10wt%)	460	53.2	57.4	6.37	No delayed

The photophysical properties of **DICzTRZ (104)** were investigated in 3 wt% mCP film. Compared to the monomer **ICzTRZ (85)**, the emission of dimer **DICzTRZ (103)** was red-shifted to 486 nm from 472 nm (**ICzTRZ, 85**), which could be ascribed to its larger conjugated donor unit. The PLQY of **DICzTRZ (103)** was increased to 61.2% from 54.7% when air was eliminated by nitrogen. Time-resolved photoluminescence behaviors were studied to reveal the emission properties of this dimer. As expected, the prompt and delayed emission in mCP were both observed with the lifetime of 28.09 ns for the prompt and 237.95 µs for the delayed components, respectively (Figure 56 and Table 18). As mentioned before, the delayed lifetime of 237.95 µs is in the range of typical TADF molecules, also confirming the TADF properties of this emitter.

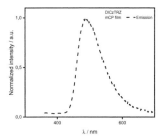 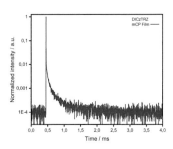

Figure 56. PL spectra (left) and photoluminescence decay curves (right) of **DICzTRZ (104)** in 3 wt% mCP film.

Table 18. Photophysical properties of the dimer **DICzTRZ (104)** and the monomer **ICzTRZ (85)** in 3 wt% mCP film.

Emitters	host	λ_{PL} thin film / nm	Φ_{PL} thin film air / %	Φ_{PL} thin film nitrogen / %	τ_p / ns	τ_d / µs
DICzTRZ (104)	mCP (3 wt%)	486	54.7	61.2	28.09	237.95
ICzTRZ (85)[a]	mCP (3 wt%)	472	53.2	57.4	4.80	113.60

[a] From Chapter 3.2.1.

In summary, a synthetic method to dimerize the carbazole- and indolocarbazole-based derivatives was developed. It was found that the reaction temperature was a crucial factor for the conversion from monomer to dimer. Especially, when using the indolocarbazole compound **ICzTRZ (85)** as substrate the dimer **DICzTRZ (104)** was obtained in a yield of 66% only when the reaction temperature was increased to 60 °C. Photophysical studies demonstrated the emission properties could be tuned between monomer and dimer. Compared to the monomer **ICzTRZ (85)**, the photoluminescence emission of the dimer **DICzTRZ (104)** was shifted to 486 nm from 472 nm (**ICzTRZ, 85**) due to the extended conjugation structure of the donor unit. The TADF nature of the monomer **ICzTRZ (85)** was succeed by **DICzTRZ (104)** with a delayed lifetime of 237.95 µs and an increased PLQY of 61.2% in 3 wt% mCP film. In contrast to the dimerization of the non-

TADF emitter **106**, emitter **DCzTRZ (103)** could not turn-on the TADF mechanism. This indicates that the conjugation through a single bond between two sp^2 carbons of the carbazole units is not a very effective conjugation extension approach, particularly when comparing with the indole-fused carbazole approach discussed in Chapter 3.2.1.

3.2.3 Methyl Torsion for TADF Emitters

As previously shown in Figure 37 (Chapter 3.2.1), fluorine atoms were introduced to the phenyl bridges of **ICzTRZ (85)** to tune the torsion and therefore modulate the conjugation between the donor and acceptor. Though the calculated ΔE_{ST} was decreased from 0.22 eV (**ICzTRZ, 85**) to 0.16 eV for **FICzTRZ (86)**, the experimental measurements showed that TADF properties of **FICzTRZ (86)** were not promising because of the absence of pronounced delayed emission. As reported, the minute ΔE_{ST} is a crucial factor to facilitate the delayed process. Therefore, a more bulky substitution, namely the methyl group, which can increase the torsion between donor and acceptor and thereby lead to smaller ΔE_{ST}, [76] was introduced to the phenyl bridge to create emitter **ICz-Me-TRZ (113)**. Meanwhile, in order to explore the influence of the electron-withdrawing strength of acceptors to TADF properties, molecule **108–112** were also designed in silico (Figure 57).

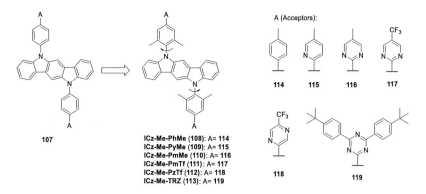

107

ICz-Me-PhMe (108): A= 114
ICz-Me-PyMe (109): A= 115
ICz-Me-PmMe (110): A= 116
ICz-Me-PmTf (111): A= 117
ICz-Me-PzTf (112): A= 118
ICz-Me-TRZ (113): A= 119

A (Acceptors):

114 **115** **116** **117**

118 **119**

Figure 57. Methyl group modified indolocarbazole derivatives **108–113**.

3.2.3.1   DFT Calculation

DFT calculations were performed using the Turbomole program package to investigate the frontier orbital distributions and energy states of the target molecules **109–113**. For **ICz-Me-PhMe (108)**, due to the absence of potential to realize TADF emission induced by no typical acceptor unit in this molecule, the calculation was not conducted for it. First, the BP86 functional and the resolution identity approach (RI) were used to optimize the geometries of the molecular structures in the ground state in the gas phase. The excitation energies were generated from time-dependent DFT (TD-DFT) methods with the optimized molecular structures. The B3LYP functional and Def2-SVP basis sets and an m4-grid numerical integration were employed to calculate the frontier orbitals and excited state energies. The initial insights on the HOMO/LUMO distribution were obtained from visualization of the molecular orbitales.

As shown in Table 19, the HOMOs of these molecules (**109–113**) are mainly located on the indolocarbazole donor unit, while the LUMOs are localized on the nitrogen-containing electron-withdrawing units and the phenyl bridges. The data (Table 20) shows that the ΔE_{ST} can be decreased from 0.22 eV (**ICzTRZ, 85**) to 0.03 eV (**ICz-Me-TRZ, 113**) due to the introduction of methyl groups to the phenyl bridges, which significantly twist the donor and acceptor units, resulting in separated frontier orbitals. When comparing the molecules with different acceptors, the ΔE_{ST} is drastically tuned by the electron-withdrawing strengths. The ΔE_{ST} of **ICz-Me-PyMe** (**109**), with a pyridyl core as acceptor, is 0.44 eV, which is too large to activate the TADF process. When the pyridine unit is replaced by the pyrimidine core, the ΔE_{ST} is decreased to 0.26 eV (**ICz-Me-PmMe, 110**). When the CF_3 group is introduced to the pyrimidine, very small ΔE_{ST} values of 0.02 eV and 0.04 eV are reached for **ICz-Me-PmTf (111)** and **ICz-Me-PzTf (112)**, respectively.

Table 19. HOMO and LUMO spatial distributions of emitters **109–113**.

Compounds	HOMO [eV]	LUMO [eV]
ICz-Me-PyMe (109)	−4.96	−1.44
ICz-Me-PmMe (110)	−4.94	−1.61
ICz-Me-PmTf (111)	−5.18	−2.27
ICz-Me-PzTf (112)	−5.33	−2.42
ICz-Me-TRZ (113)	−5.06	−1.99

Table 20. DFT Calculations of **109–113**.

Compounds	HOMO/LUMO[a] [eV]	ΔE^a [eV]	S_1 [eV]	T_1 [eV]	ΔE_{ST} [eV]
ICz-Me-PyMe (109)	−4.96/−1.44	3.52	3.04	2.60	0.44
ICz-Me-PmMe (110)	−4.94/−1.61	3.33	2.86	2.60	0.26
ICz-Me-PmTf (111)	−5.18−2.27	2.91	2.45	2.47	0.02
ICz-Me-PzTf (112)	−5.33/−2.42	2.91	2.49	2.45	0.04
ICz-Me-TRZ (113)	−5.06/−1.99	3.07	2.64	2.61	0.03

3.2.3.2 Synthesis

Through retrosynthetic analysis, the building block **124** or **125** was identified as the key intermediate to obtain the target compounds through palladium catalyzed cross-coupling. For this, nucleophilic aromatic substitution (S_NAr) reactions of 5-bromo-2-fluoro-1,3-dimethylbenzene with **ICz (76)** under various common reaction conditions were tested (Table 21, entries 1 and 2). Unfortunately, no desired product (**124**) was obtained, as electron withdrawing fluoroarenes are crucial to an efficient S_NAr reactivity. To circumvent this, the Buchwald-Hartwig cross-coupling was conducted using iodo- (Table 21, entry 3) or bromo-arene substrates (Table 21, entry 4). However, the target building block could still not be accessed which is presumably due to the large steric bulk generated by the two methyl groups at the *ortho* position to the halogenides.

Table 21. Nucleophilic aromatic substitution and Buchwald-Hartwig cross-coupling to synthesize the building blocks **124** or **125**.

74

120: X=Br, Y= F
121: X=Br, Y= F
122: X=Br, Y= I
123: X=Cl, Y= Br

124: X=Br
125: X=Cl

Entry	X, Y	Conditions	124 or 125/Yield
1	X=Br, Y=F	Cs$_2$CO$_3$, DMF, Argon, 150 °C, 12 h	No target product
2	X=Br, Y=F	NaH, DMF, Argon, 100°C, 12 h	No target product
3	X=Br, Y=I	Pd(OAc)$_2$, P(^{t}Bu)$_3$BF$_4$, NaO^{t}Bu, DMF, Argon, 100°C, 12 h	No target product
4	X=Cl, Y=Br	Pd(OAc)$_2$, P(^{t}Bu)$_3$BF$_4$, NaO^{t}Bu, PhMe, Argon, 100°C, 12 h	No target product

Therefore, a new method was designed to synthesize the key intermediate **125**. As shown in Scheme 18, the successful construction of N-functionalized indolo[3,2-*b*]carbazole relies on the preeminent six-fold Pd-catalyzed Buchwald-Hartwig cross-coupling of the tetrabromo terphenyl precursor **127** with the sterically hindered 4-chloro-2,6-dimethylaniline whereby two rings are formed in the process.[142] For the synthesis of coupling product **125**, the use of Pd$_2$(dba)$_3$ in combination with ^{t}Bu$_2$PPh as a catalyst system and NaO^{t}Bu as a base in a refluxing PhMe/HO^{t}Bu solvent mixture proved to be more effective in terms of higher yield of 74%. Precursor **127** is formed by Suzuki-Miyaura cross-coupling of an aromatic 2-bromo-phenylboronic acid and 1,4-dibromo-2,5-diiodobenzene.[143] The Pd-catalyzed carbon-carbon bond forming reaction using Pd(PPh$_3$)$_4$ as catalyst, Ag$_2$CO$_3$ as base and THF/H$_2$O solvent mixture, proceeded with predominantly excellent regioselectivity in favor of the iodo- groups with 78% yield.

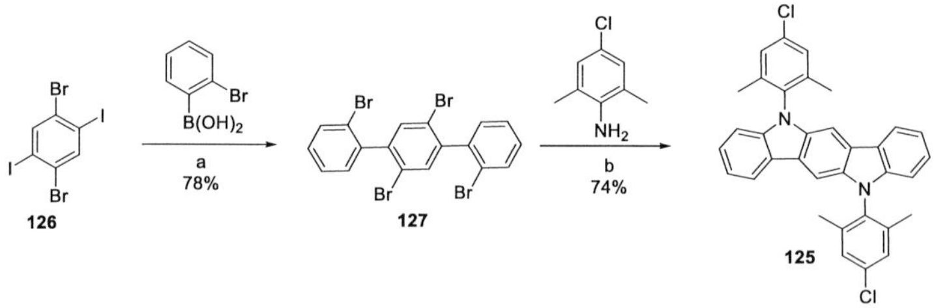

Scheme 18. Synthesis of functionalized indolo[3,2-*b*]carbazole derivatives. a) Pd(PPh$_3$)$_4$, Ag$_2$CO$_3$, THF/H$_2$O, Argon, 80°C, 12 h. b) Pd$_2$(dba)$_3$, ^{t}Bu$_2$PPh, NaO^{t}Bu, PhMe/HO^{t}Bu, Argon, 100°C, 1.5 h.

In the key intermediate **125**, the two methyl groups at the *ortho* position to the indolocarbazole donor are intended to induce a significant twist between the donor and acceptor, while the two chloride functions at the peripheries can be exploited under borylation reaction conditions with bis(pinacolato)diboron (B_2pin_2) to prepare the boronic ester **128**, which is the key strategic intermediate for the addition of a variety of acceptors by Suzuki-Miyaura cross-coupling to obtain a set of symmetrical acceptor-donor-acceptor emitters **108–113** (Scheme 19). The reaction conditions were optimized employing $Pd(PPh_3)_4$ and K_2CO_3 as catalyst system in refluxing PhMe/EtOH solvent mixture at 90 °C for 12 h, providing the target compounds in 65-80% yields.

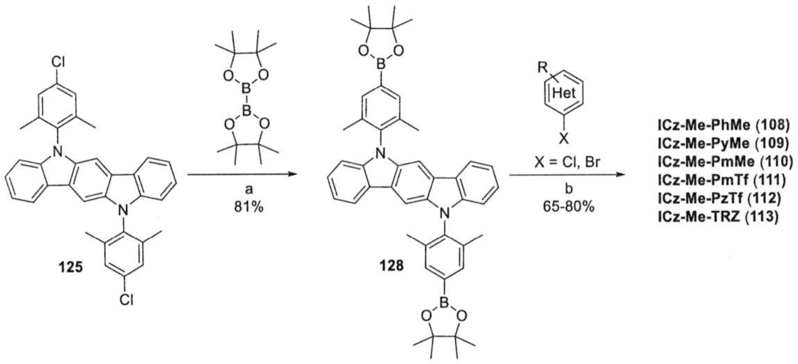

ICz-Me-PhMe (108)
ICz-Me-PyMe (109)
ICz-Me-PmMe (110)
ICz-Me-PmTf (111)
ICz-Me-PzTf (112)
ICz-Me-TRZ (113)

Scheme 19. Suzuki-Miyaura cross-coupling to indolo-carbazole-based emitters **108–113** with increased twist. a) $Pd_2(dba)_3$, Xphos, KOAc, Dioxane, Argon, 110°C, 12 h. b) $Pd(PPh_3)_4$, K_2CO_3, PhMe/EtOH, 90°C, 12h.

These emitters were fully characterized by NMR, mass spectra (MS and HRMS) and IR. Molecular structures of compounds **ICz-Me-PyMe (109), ICz-Me-PmMe (110), ICz-Me-PmTf (111) and ICz-Me-PzTf (112)** were further confirmed by single crystal X-ray structure analysis (Figure 58). These molecules possess a twisted geometry between the phenyl bridge and indolocarbazole unit with torsion angles at around 80 degrees, which can lead to minute ΔE_{ST} values as predicted by DFT calculation. In addition, the torsion angles vary a bit as they are influenced by the different nitrogen-containing acceptor groups. For example, the angle of **ICz-Me-PmTf (111)** amounts to 83.22°, while it decreases to 75.05 ° for **ICz-Me-PzTf (112)** when the pyrimidyl unit is replaced by pyrazine.

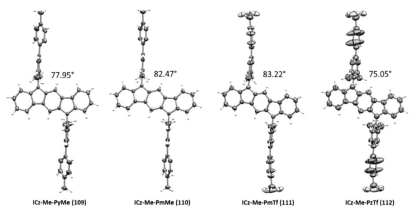

Figure 58. Molecular structures of the differently functionalized twisted molecules **ICz-Me-PyMe (109)**, ICz-Me-PmMe **(110)**, **ICz-Me-PmTf (111) and ICz-Me-PzTf (112)** (from left to right) and dihedral angles between the phenyl bridge and indolocarbozole unit.

3.2.3.3   Photophysics

The photophysical and electrochemical properties of these emitters were performed *via* collaboration with cynora GmbH. All photophysical and electrochemical measurements in this chapter were provided by Dr. Stefan Seifermann and Nico-Patrick Thöbes.

The photophysical properties of the promising emitters **ICz-Me-PmTf (111)**, **ICz-Me-PzTf (112)** and **ICz-Me-TRZ (113)**, which possess minute ΔE_{ST} (0.02–0.04 eV), were investigated as films doped in 10 wt% PMMA. In order to confirm the non-TADF nature of **ICz-Me-PhMe (108)**, the photophysical properties were also studied. As shown in Figure 59, the photoluminescence emission maximum of **ICz-Me-PhMe (108)** is at 412 nm accompanied with a shoulder peak at 436 nm. The emission of **ICz-Me-PmTf (111)**, **ICz-Me-PzTf (112)** and **ICz-Me-TRZ (113)** are all red-shifted between 479 nm and 503 nm. For the quantum efficiency, **ICz-Me-PzTf (112)** shows the best performance with a PLQY of 53%. For **ICz-Me-TRZ (113)**, which emits in the sky-blue region with 478 nm, the PLQY is decreased to only 33%.

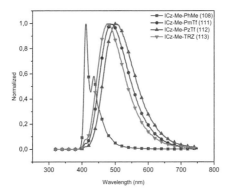

Figure 59. PL spectra of **ICz-Me-PhMe (108)**, **ICz-Me-PmTf (111)**, **ICz-Me-PzTf (112)** and **ICz-Me-TRZ (113)** in 10 wt% PMMA.

Time-resolved photoluminescence spectra were studied to reveal the emission properties of these emitters. For **ICz-Me-PmTf (111)**, **ICz-Me-PzTf (112)** and **ICz-Me-TRZ (113)**, the experimental results demonstrate that they are TADF emitters as they possess delayed lifetime components of 346 µs, 169 µs and 217 µs for **ICz-Me-PmTf (111)**, **ICz-Me-PzTf (112)** and **ICz-Me-TRZ (113)**, respectively (Figure 60 and Table 22). This is consistent with the theoretical predictions. Meanwhile, **ICz-Me-PhMe (108)** only shows prompt fluorescence with a lifetime of 8.7 ns. This is due to the very weak electron-withdrawing strength of the 4-toluyl unit.

Table 22. Photophysical properties of **ICz-Me-PhMe (108)**, **ICz-Me-PmTf (111)**, **ICz-Me-PzTf (112)** and **ICz-Me-TRZ (113)** in 10 wt% doped films in PMMA at 298 K.

Compounds	λ_{max}[nm]	FWHM [eV]	PLQY[%]	Delayed lifetime/τ
ICz-Me-PhMe (108)	412 (436)	-	51	$-^a$
ICz-Me-PmTf (111)	489	0,52	49	346 µs
ICz-Me-PzTf (112)	503	0,51	53	169 µs
ICz-Me-TRZ (113)	478	0,50	33	217 µs

^{a} only prompt lifetime of 8.7 ns.

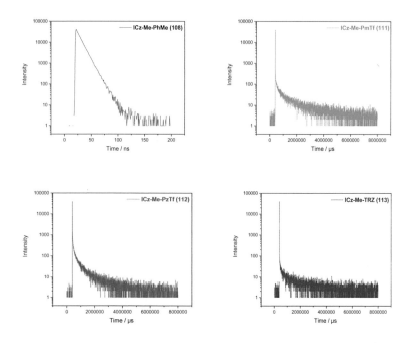

Figure 60. Photoluminescence decay curves of **ICz-Me-PhMe (108)**, **ICz-Me-PmTf (111)**, **ICz-Me-PzTf (112)** and **ICz-Me-TRZ (113)**.

In summary, a series of symmetrical acceptor-substituted novel indolo[3,2-*b*]carbazoles with increased torsion at around 80° were designed and synthesized by a four-step procedure which involved a one-pot sequential Buchwald-Hartwig amination/cyclization process, followed by a Pd-catalyzed step-wise borylation/Suzuki–Miyaura cross-coupling protocol. By increasing the steric hindrance between donor and acceptor moieties, the spatial overlap of HOMO and LUMO can be effectively broken to obtain small singlet-triplet energy splitting (ΔE_{ST}) as complemented by theoretical calculations. Experimental results indicate that the photophysical properties of the indolo[3,2-*b*]carbazole-based molecules can be effectively tuned while maintaining the thermally activated delayed fluorescence (TADF) process. Furthermore, varying the electron-withdrawing strength of the acceptor moieties in a sterically demanding twisted arrangement shows an effect on TADF emission with the color ranging from 478 nm to 503 nm.

3.3 [2.2]Paracyclophane-Based Donors for TADF Emitters

The [2.2]paracyclophane (**PCP, 129**) is a unique molecule that contains two slightly bent benzene rings which are connected by ethylene bridges (Figure 61).[81, 144-148] Due to the spatial stacked configuration and short distance of 2.78–3.09 Å between the two benzene units, which is smaller than the van der Waals distance between the layers of graphite (3.35 Å), a strong transannular electronic communication can take place. This electronic interaction can be extended through at least ten stacked layers bridged by [2.2]paracyclophane units, which clearly shows the efficiency of through-space interactions within the benzene decks of the [2.2]paracyclophane.[84-85]

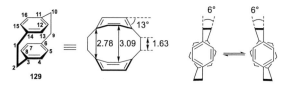

Figure 61. Structural parameters of [2.2]paracyclophane.

In this regard, TADF emitters based on a [2.2]paracyclophane bridging structures have been realized showing through-space charge transfer. As mentioned in Chapter 1.2.2, Bräse *et al.* firstly reported two TADF molecules *via* combining the donor and acceptor units on the different decks of [2.2]paracyclophane. Photoluminescence quantum yields of 60% for ***trans*-Bz-PCP-TPA (19)** and 45% for ***cis*-Bz-PCP-TPA (20)** in deoxygenated solution were obtained.[87] Small ΔE_{ST} values of 0.17 eV (***trans*-Bz-PCP-TPA, 19**) and 0.04 eV (***cis*-Bz-PCP-TPA, 20**) suggest their potential for TADF emission. Temperature dependence of delayed lifetimes were investigated to examine the TADF properties. As expected, short delayed lifetimes of 3.6 µs for ***trans*-Bz-PCP-TPA (19)** and 1.8 µs for ***cis*-Bz-PCP-TPA (20)** were achieved, confirming the TADF emission process (Figure 62).

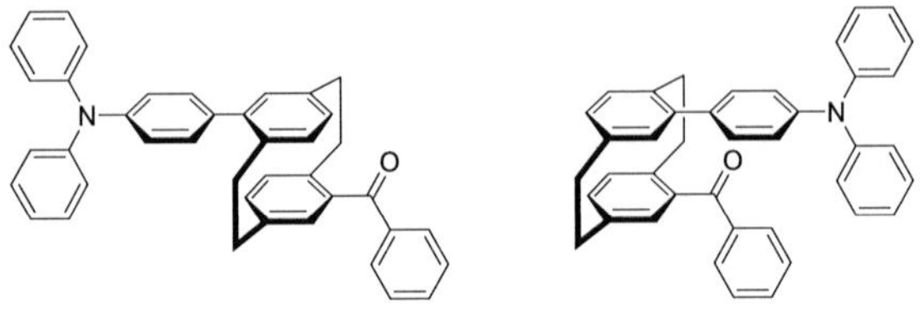

19
trans-Bz-PCP-TPA

20
cis-Bz-PCP-TPA

Figure 62. Chemical structures of [2.2]paracyclophane-based TADF emitters, *trans*-**Bz-PCP-TPA** (**19**) and *cis*-**Bz-PCP-TPA** (**20**).

3.3.1 Carbazolophane-Based TADF Emitters

Besides using the [2.2]paracyclophane scaffold as a bridge for the through-space charge transfer between the donor and acceptor moieties, Bräse *et al.* have also extended the [2.2]paracyclophane skeleton to the donor unit carbazolophane (**Czp, 130,** Scheme 63), which possesses additional torsion enhanced by adjacent ethylene bridge, an enlarged donor volume for a more dissipated HOMO induced by through-space interactions and planar chirality.[78]

Based on this donor, the TADF emitter **CzpPhTRZ** (**131**) with a small ΔE_{ST} of 0.16 eV and PLQY of 70% in doped film was obtained. The TADF properties were confirmed by time dependent emission lifetime measurements showing a delayed lifetime of 65 µs, and the typical temperature dependence of delayed emission. Furthermore, OLED properties with a high maximum EQE of 17% and of 12% at 100 cd m^{-2} were achieved. However, compared with the analogous emitter **CzPhTRZ** (**8**) based on a carbazole donor, which does not possess TADF properties, the emission wavelength of **CzpPhTRZ** (**131**) was red-shifted from deep blue (446 nm) to sky-blue (480 nm).

130
Czp

131
CzpPhTRZ

TADF turn-on

8
CzPhTRZ

Figure 63. Chemical structures of **Czp** (**130**), TADF emitter **CzpPhTRZ** (**131**), non-TADF emitter **CzPhTRZ** (**8**).

As deep blue emission is very important and desired for practical applications of OLEDs, the focus of this project lays in tuning the parent emitter structure **CzpPhTRZ** (**131**) in such a way that bluer emission is reached through the introduction of electron-withdrawing groups, while the TADF property remains.

3.3.1.1 Molecular Design and DFT Calculation

According to literature,[15, 149-150] electron-withdrawing groups deepen the HOMO level, leading to a larger optical gap and then resulting in bluer emission. Therefore, electron-withdrawing groups such as cyano (CN), trifluoromethyl (CF_3) and benzoyl (Bz) were introduced to the donor part (**Czp**) of the parent emitter, **CzpPhTRZ** (**131**). This is done conveniently at the *para*-position of the nitrogen atom, which is the most reactive site for electrophilic functionalization such as bromination. The structures of the parent emitter, three target molecules and two molecules obtained unexpectedly from the synthesis of the benzoyl derivative are shown in Figure 64.

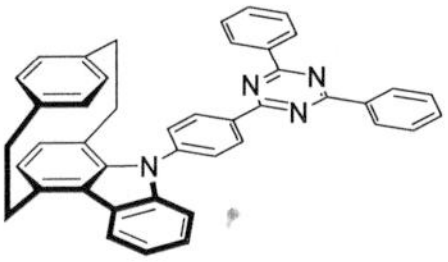

131
CzpPhTRZ

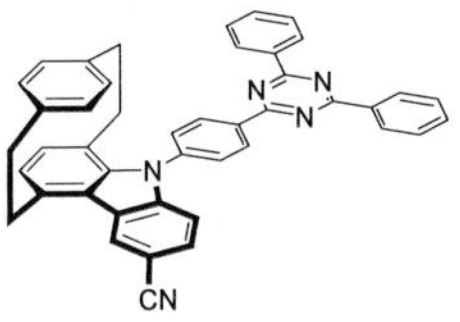

132
CyCzpPhTRZ

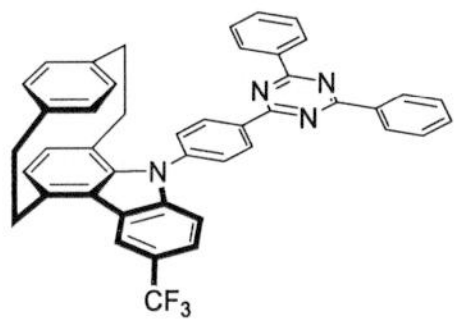

133
TfCzpPhTRZ

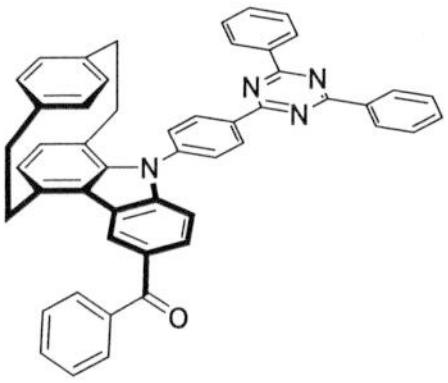

134
BzCzpPhTRZ-1

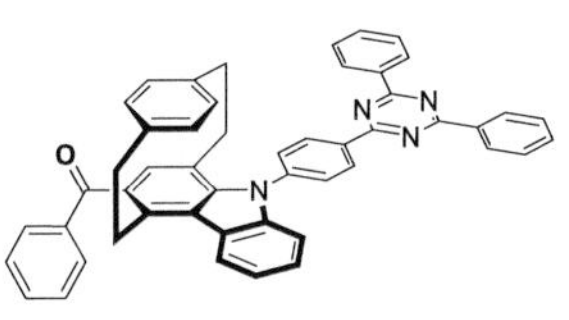

135
BzCzpPhTRZ-2

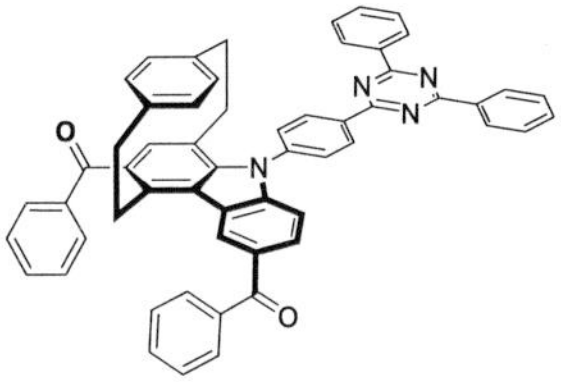

136
BzCzpPhTRZ-3

Figure 64. Chemical structures of **Czp**-based molecules **131–136**.

DFT calculations were performed employing the Gaussian 09 revision D.018 suite to investigate the frontier orbital densities and energy states of these molecules. First, the PBE0 functional with the standard Pople 6-31G (d,p) basis set was used to optimize the geometries of the molecular structures in the ground state in gas phase. Next, time-dependent DFT calculations were performed with the Tamm–Dancoff approximation (TDA) based on the optimized molecular structures. The GaussView 5.0 software was employed to visualize the molecular orbitals and provide initial insights on the nature of the modified properties induced by different donors. The frontier orbitals (HOMO/LUMO) and energy levels of these molecules are shown in Figure 65, Figure 66 and Table 23.

The DFT calculations suggest that the frontier orbitals and energy states are effectively tuned *via* the introduction of electron-withdrawing groups. Comparing the HOMOs of the derivatives with the parent emitter **CzpPhTRZ (131)** (–5.54 eV), these were deepened to ⁻5.91 eV (**CyCzpPhTRZ, 132**), ⁻5.78 eV (**TfCzpPhTRZ, 133**) and ⁻5.74 eV (**BzCzpPhTRZ-1, 134**) (Figure 65 and Table 23). Though the LUMO levels of the derivatives are slightly decreased, the energy gaps of these three molecules are significantly increased due to the drastically reduced HOMO levels. For the excited singlet states, which is the decisive factor for the final emission color, all these three derivatives are increased from 3.12 eV (**CzpPhTRZ, 131**) to 3.27 eV (**CyCzpPhTRZ, 132**), 3.23 eV (**TfCzpPhTRZ, 133**) and 3.20 eV (**BzCzpPhTRZ-1, 134**), respectively. As the torsion angles are similar between the **Czp** donor and triazine acceptor, and the electronic volume of the substituents are small compared with the **Czp** unit, the values of ΔE_{ST} of the new designed molecules **CyCzpPhTRZ (132)** (0.31 eV) and **TfCzpPhTRZ (133)** (0.31 eV) are identical to **CzpPhTRZ (131)** (0.31 eV), which suggests that these three molecules will show equally promising TADF processes. The oscillator strengths are slightly tuned between 0.3185 and 0.3886 because of the electron withdrawing strength and therefore modulation of the spatial distribution of the HOMO orbital for different substituents.

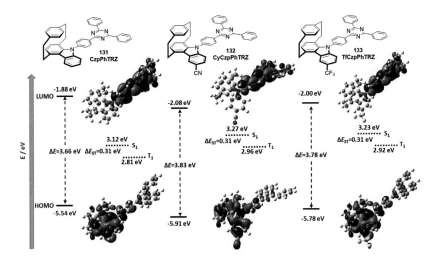

Figure 65. DFT calculations of **CzpPhTRZ** (**131**, left), **CyCzpPhTRZ** (**132,** central) and **TfCzpPhTRZ** (**133,** right). Results provided by Dr. Eduard Spuling.

The DFT calculations were also applied to another two molecules, which were obtained during the synthesis of **BzCzpPhTRZ-1 (134)** and as additional products (*vide infra*). The **BzCzpPhTRZ-2 (135)** is the isomer of **BzCzpPhTRZ-1 (134)**, which is mono-benzoylated at the PCP skeleton (Figure 66 and Table 23). **BzCzpPhTRZ-3 (136)** is the di-benzoylated product with the substituents at both *para* positions of the nitrogen atom of **Czp**. DFT calculations show that different substitution positions and numbers of benzoyl groups also tune the energy states. Compared with the mono-substituted products, the bis-substituted molecule possesses a higher lying HOMO level and therefore higher lying excited singlet state due to the stronger electron-withdrawing ability of two benzoyl groups.

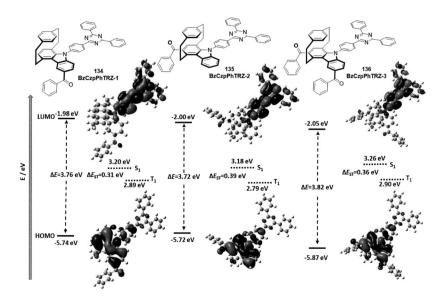

Figure 66. DFT calculations of **BzCzpPhTRZ-1** (**134**, left), **BzCzpPhTRZ-2** (**135,** central) and **BzCzpPhTRZ-3** (**136**, right). Results were provided by Dr. Eduard Spuling.

Table 23. DFT Calculations of **Czp**-based emitters **131–136**.

Compounds	HOMO/LUMO [eV]	ΔE [eV]	f	S_1 [eV]	T_1 [eV]	ΔE_{ST} [eV]
CzpPhTRZ (131)	−5.54/−1.88	3.66	0.3420	3.12	2.81	0.31
CyCzpPhTRZ (132)	−5.91/−2.08	3.83	0.3185	3.27	2.96	0.31
TfCzpPhTRZ (133)	−5.78/−2.00	3.78	0.3250	3.23	2.92	0.31
BzCzpPhTRZ-1 (134)	−5.74/−1.98	3.76	0.3886	3.20	2.89	0.31
BzCzpPhTRZ-2 (135)	−5.72/−2.00	3.72	0.3814	3.18	2.79	0.39
BzCzpPhTRZ-3 (136)	−5.87/−2.05	3.82	0.3953	3.26	2.90	0.36

3.3.1.2   Synthesis Carbazolophane-Based Emitters

The synthetic access to **Czp** was reported by Bolm *et al.* in 2011 through a two-step procedure.[151] First, the intermediate, secondary amine **138,** is obtained *via* Hartwig-Buchwald cross-coupling of the racemic bromo-[2.2]paracyclophane **137** to aniline. Next, a Pd-catalyzed oxidative cyclization is used to generate the novel donor **Czp (137)**. Recently, the multigram preparation of **Czp** was successfully conducted by Bräse *et al.* using a modified procedure.[18] However, the yield of the oxidative cyclization process was only 26% when the reaction was scaled up to multigram level (2.07 g), resulting in the overall yield of 22% by this procedure (Scheme 20).

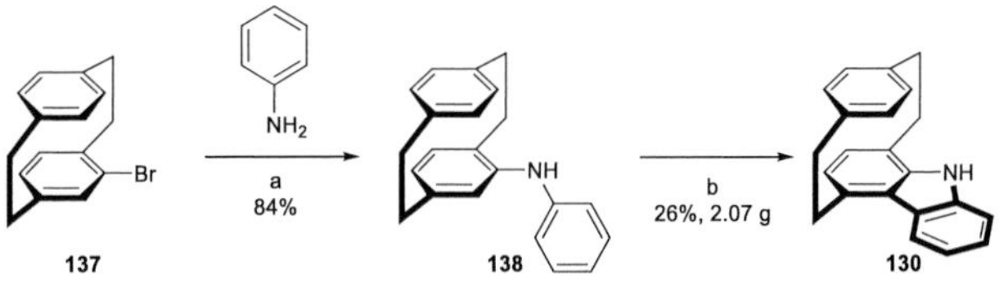

Scheme 20. Synthetic route to **Czp** reported by Bolm and Bräse. a) Pd$_2$(dba)$_3$, SPhos, KO^{t}Bu,toluene, Argon, 110 °C, 16 h. b) Pd(OAc)$_2$, PivOH, air, HOAc, 120 °C, 12 h.

Buchwald and co-workers reported an alternative approach to **Czp** *via* the *ortho*-chlorinated phenylamino substituted [2.2]paracyclophane **140** while the chlorine atoms acts as the directing group for the oxidative cyclization process.[152] Although the yield of the final cyclization is increased to 58%, it is a three-step procedure with an overall yield of 37% in 1.30 g scale (Scheme 21).

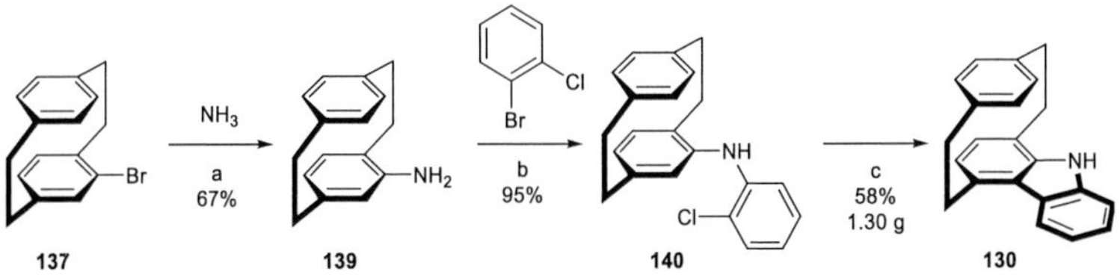

Scheme 21. Synthesis of **Czp (130)** reported by Buchwald *et al.* a) Pd-^{t}BuDavesPhos-OMs, NaO^{t}Bu, 1,4-dioxane, 80 °C, 16 h. b) Pd-DavesPhos-OMs, NaO^{t}Bu, 1,4-dioxane, 80 °C, 3 h. c) Pd-PCy$_3$-OMs, PivOH, K$_2$CO$_3$, DMA, 110 °C, 16 h.

Therefore, it was necessary to find a more robust method for the multigram synthesis of **Czp** by combining the methods mentioned above.[153-154] In this part, a new procedure was developed to synthesize **Czp**. By using the commercially available 2-chloroaniline, the key intermediate **140** was obtained in only one step with a yield of 76%. This intermediate was oxidatively cyclized under the modified conditions as reported by Buchwald *et al.* to prepare the final product in 60% yield with a 3.20 g scale (Scheme 22). The overall yield of this method is 46% in two steps, demonstrating the improved accessibility to the key intermediate **Czp** on the multigram scale.

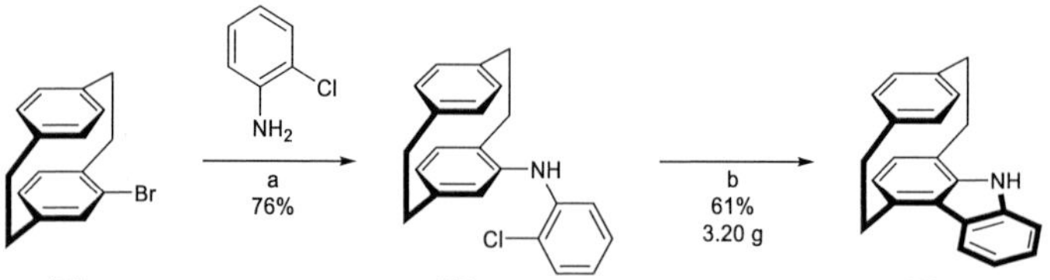

Scheme 22. Synthesis of **Czp** (**130**). a) Pd(OAc)$_2$, P^{t}Bu$_3$HBF$_4$, NaO^{t}Bu, toluene, Argon, 100 °C, 12 h. b) Pd$_2$(dba)$_3$, X-Phos, PivOH, HOAc, K$_2$CO$_3$, DMA, Argon, 110 °C, 12 h.

As a first functionalization attempt, the bromination of **Czp** was explored because of the chemical versatility of bromoarene groups. Bromination with NBS as bromine source at room temperature was conducted to yield the brominated species **141** with the bromine selectively in the non-cyclophanic arenein with a very good yield of 83% (Scheme 23).[18]

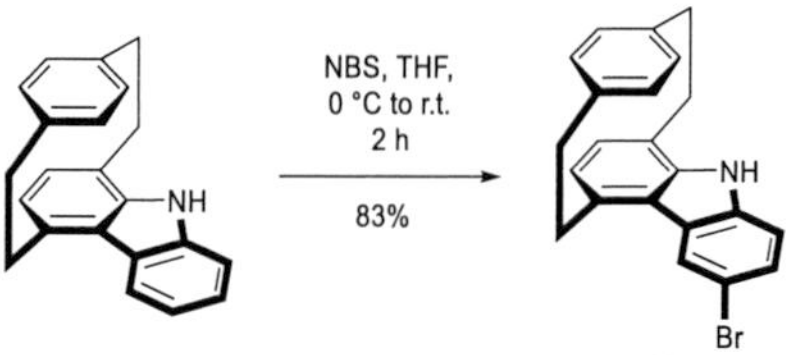

Scheme 23. Bromination of **Czp** with NBS as the bromine source.

Synthesis of CyCzpPhTRZ (132)

From the brominated intermediate **141**, the desired and electron withdrawing cyano group can be readily obtained by the Rosenmund-von Braun reaction. This was reported to work smoothly when the molecular skeleton is a carbazole unit.[155-158] Similar conditions were used to prepare the cyanated **Czp** species **142** in a very good yield of 89% (Scheme 24).

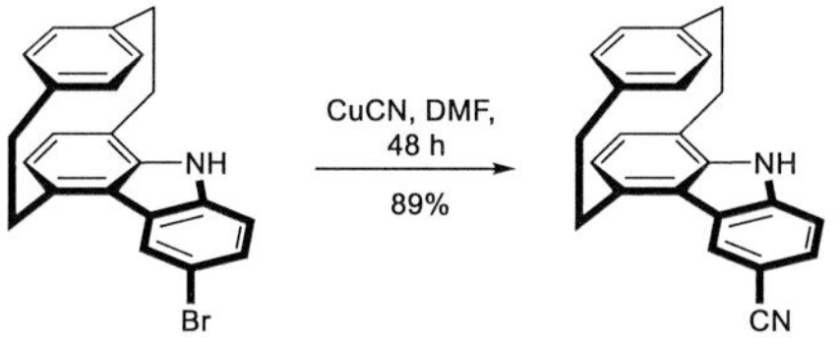

Scheme 24. Synthesis of cyano-substituted **Czp**.

Afterwards, donor **142** and triazine acceptor **98** were linked together *via* a nucleophilic aromatic substitution using tripotassium phosphate as base in DMSO at 120 °C (Scheme 25). The final emitter **CyCzpPhTRZ (132)** was obtained as a colorless solid and strong blue luminescence was observed under 366 nm UV light.

Scheme 25. Synthesis of **CyCzpPhTRZ (132)**.

Synthesis of TfCzpPhTRZ (133)

For introducing the trifluoromethyl group to the **Czp** donor, the brominated intermediate **141** was treated with sodium trifluoroacetate in the presence of copper(I) iodide at 160 °C in DMA (Scheme 26).[159-160] After 12 hours no conversion was detected by TLC analysis. Extending the reaction time to 48 hours still did not show any conversion of the starting material.

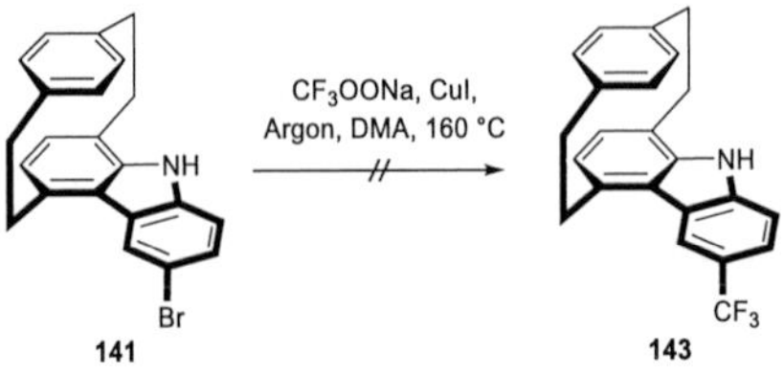

Scheme 26. Synthesis of trifluoromethyl-functionalized donor **143** from bromo derivative **141**.

As these harsh reaction conditions to obtain trifluoromethyl group from bromine-containing compound did not give any product, a *de novo* synthesis was conducted under the same reaction conditions as for **Czp** starting from a 4-trifluoromethylated aniline derivative coupled to bromo[2.2]paracyclophane and subsequent cyclization of intermediate **144**. Intermediate **144** was obtained in 55% yield. Following Pd-catalyzed oxidative cyclization gave the trifluoromethyl-functionalized donor unit **143** in 62% yield (Scheme 27).

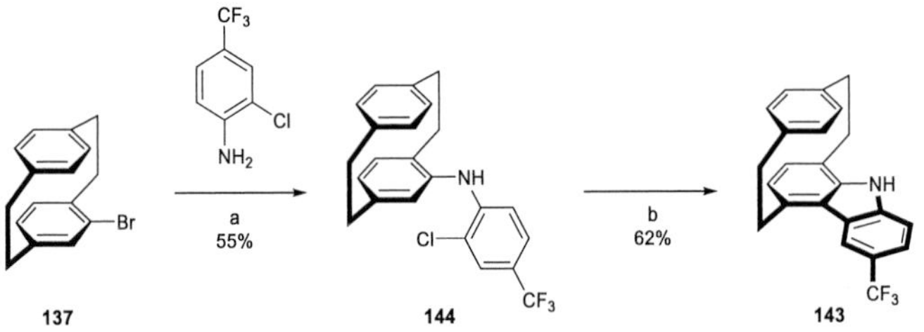

Scheme 27. Synthetic route to trifluoromethyl-functionalized donor **143**. a) Pd$_2$(dba)$_3$, X-Phos, NaO^{t}Bu, toluene, Argon, 100 °C, 12 h. b) Pd$_2$(dba)$_3$, XPhos, PivOH, HOAc, K$_2$CO$_3$, DMA, Argon, 110 °C, 12 h.

Lastly, the prepared donor **143** and triazine-based acceptor **98** were linked together *via* a nucleophilic aromatic substitution using tripotassium phosphate as base in DMSO at 120 °C yielding the final emitter in 39% as a white solid and strong blue luminescence was observed under UV excitation (Scheme 28).

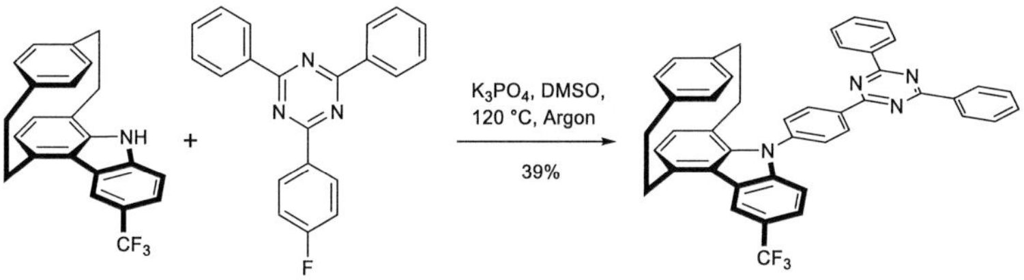

Scheme 28. Synthesis of **TfCzpPhTRZ (133)**.

Synthesis of Benzoyl-functionalized Emitters 134–136

For the synthesis of the benzoyl derivative **145** starting from the bromo intermediate **141**, the classic organometallic approach was tested first. Considering the competition between the bromide-metal exchange and deprotonation of **141**, the acceptor part was introduced to **141** before the following organometallic process. By using the condition shown in Scheme 27, the nucleophilic aromatic substitution reaction of bromo derivate **141** and fluoro-triazine **98** gave intermediate **145** in 64% yield (Scheme 29).

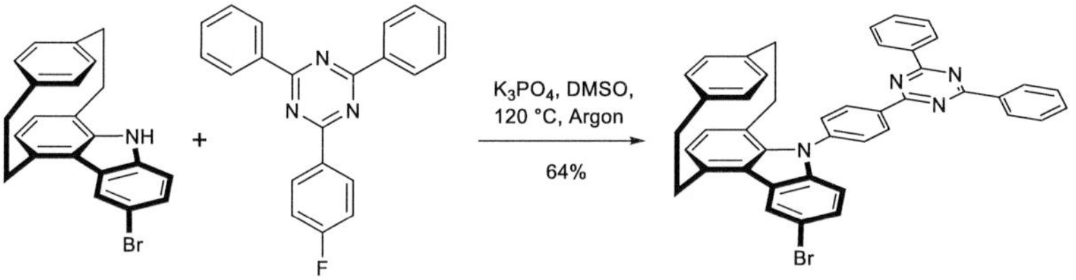

Scheme 29. Synthesis of intermediate **145**.

Then, intermediate **145** was treated with an excess *n*-butyllithium at low temperatures followed by addition of benzoyl chloride (Scheme 30). Unfortunately, no desired benzoyl product was obtained and only starting material and unidentified side products were observed. Another approach *via* organometallic chemistry is the use of organomagnesium reagents. But, based on the poor solubility of **145** in common solvents for organometallic procedures, such as ether or tetrahydrofuran, this method was not tested.

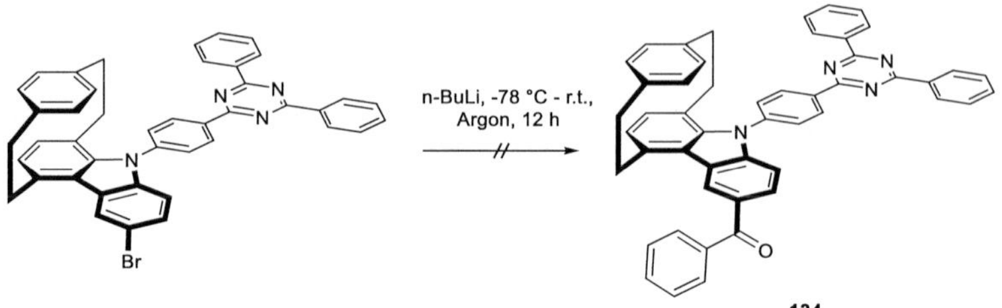

145 **134** **BzCzpPhTRZ-1**

Scheme 30. Synthetic route to **134** through organometallic process.

Lewis acid catalyzed Friedel-Crafts acylation allows the introduction of a benzoyl group *via* electrophilic aromatic substitution using benzoyl chloride directly to aromatic hydrocarbon moieties. For carbazole derivatives, the preferential position of the benzoylation is the 3-position, *para* to the nitrogen, as reported by literature (Scheme 31).[161-165]

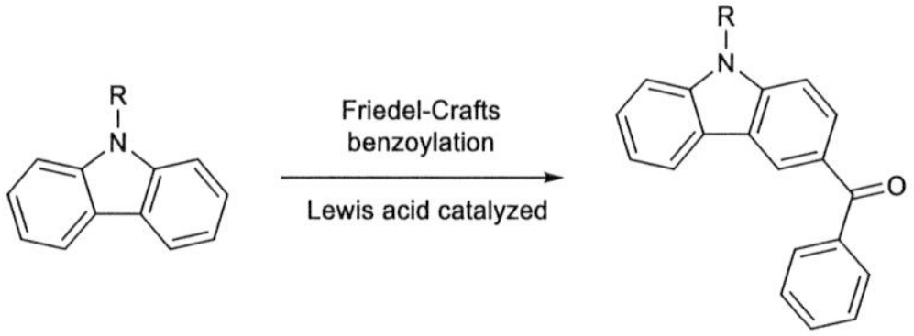

146 **147**

Scheme 31. Lewis acid catalyzed Friedel-Crafts benzoylation for carbazole-based compounds.

Again in order to avoid potential side reactions at the N-H function of **Czp (130)**, the acceptor part was reacted with **Czp (130)** to generate **CzpPhTRZ (131)**, the parent molecule of these **Czp**-based compounds, before the following benzoylation. The parent emitter **CzpPhTRZ (131)** was obtained by the identical nucleophilic aromatic substitution protocol, yielding the product in very good yield of 86% (Scheme 32).

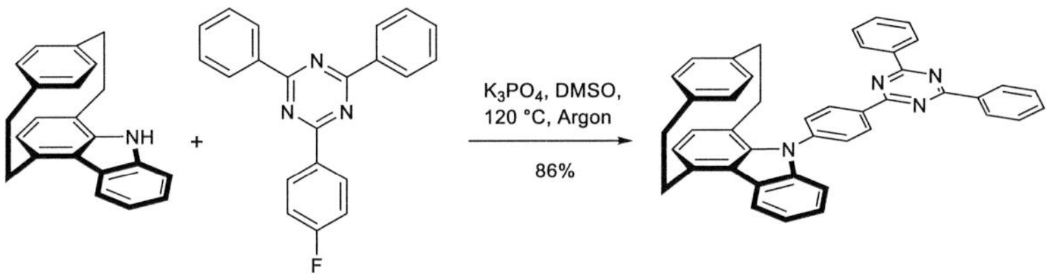

Scheme 32. Synthesis of the parent molecule **CzpPhTRZ (131)**.

Although numerous aromatic groups are present in intermediate **CzpPhTRZ (131)**, the electron donating strength of the **Czp** and electron withdrawing strength of the triazine acceptor unit are expected to guide the electrophilic benzoylation selectively to the desired position on the donor unit. First, 1.10 equivalents of benzoyl chloride were reacted with **CzpPhTRZ (131)** in the presence of AlCl$_3$ catalyst. After workup of the reaction mixture, two isomers were obtained with the benzoyl group at the different position of **Czp (130)** donor unit. The structures of **BzCzpPhTRZ-1 (134)** and **BzCzpPhTRZ-2 (135)**, were confirmed by NMR and mass analyses. Interestingly, when compared to the electrophilic bromination using NBS, where the bromination selectively occurs on the non-cyclophanic *para* position of the carbazole subunit of **Czp (130)**, the benzoylation appears to favor the cyclophanic and sterically more shielded *para* position. When using 1.10 equivalent of benzoylchloride, a mixture of **BzCzpPhTRZ-1 (134)** and **BzCzpPhTRZ-2 (135)** is isolated with 5% and 57% yield, respectively. Therefore, the selectivity is not explicit, but the most reactive position is the cyclophanic *para* position. In addition, traces of the dibenzoylated product **BzCzpPhTRZ-3 (136)** were detected, while the starting material appeared not to be fully consumed after 24 h. This prompted an increase of benzoyl chloride equivalents to 2.20. In this reaction, the starting material was consumed completely and with the dibenzoylated product **BzCzpPhTRZ-3 (136)** could be isolated in 24% yield, while the yields of the mono-benzoylated products decreased to 3% and 50%, respectively (Table 24).

Table 24. Friedel-Crafts benzoylation for benzoyl-functionalized emitters **134–136**.

Entry	148/[equiv.]	BzCzpPhTRZ-1	BzCzpPhTRZ-2	BzCzpPhTRZ-3
1	1.10	5%	57%	Only traces
2	2.20	3%	50%	24%

Due to the difficulty to obtain **BzCzpPhTRZ-1 (134)** in sufficient amounts because of the very low yield, further photophysical measurements were only conducted for the other two benzoyl based emitters.

3.3.1.3   Photophysics

UV-Vis absorption and photoluminescence behaviors of **CyCzpPhTRZ (132)**, **TfCzpPhTRZ (133)**, **BzCzpPhTRZ–2 (135)** and **BzCzpPhTRZ–3 (136)** in neat film and 10 wt% doped in DPEPO were measured (Figure 67). The absorption peaks of these four emitters between 350 to 400 nm can be ascribed to local transitions in the **Czp (130)** donor and triazine-based acceptor moieties. Compared with the parent emitter **CzpPhTRZ (131)**, the emission of these four emitters is drastically blue-shifted from 476 nm to 448–460 nm, which confirms that the introduction of electron-withdrawing groups to the **Czp (130)** donor is an efficient method to get bluer emitter. For the PLQY in neat films, **CyCzpPhTRZ (132)** and **TfCzpPhTRZ (133)** show comparable values of 36% and 32% to **CzpPhTRZ (131)** of 40%. However, the PLQY of **BzCzpPhTRZ–2 (135)** and **BzCzpPhTRZ–3 (136)** is dropped to only 7% and 3%, respectively. Furthermore, in 10 wt% DPEPO films, theses four emitters also exhibited deep blue emission with λ_{PL} of 443–454 nm, while **CzpPhTRZ (131)** presents sky-blue emission at λ_{PL} of 482 nm. Though the PLQY of **TfCzpPhTRZ (133)** is not as high as **CzpPhTRZ (131)** (69%), it can still reach up to 49%. Unfortunately, **BzCzpPhTRZ–2 (135)** and **BzCzpPhTRZ–3 (136)** show low values of 13% and 5%.

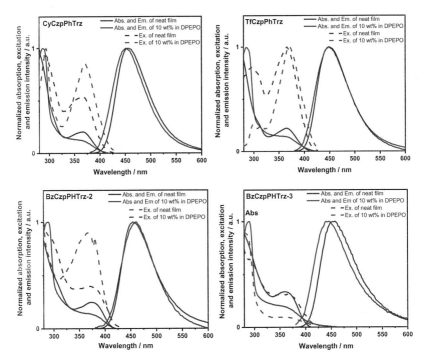

Figure 67. UV-Vis absorption and PL spectra of **CyCzpPhTRZ (132)**, **TfCzpPhTRZ (133)**, **BzCzpPhTRZ–2** (135) and **BzCzpPhTRZ–3 (136)** in neat thin film and doped in host materials. Results were provided by Dr. Abhishek Kumar Gupta.

Table 25. Photophysical properties of **CzpPhTRZ (131)**, **CyCzpPhTRZ (132)**, **TfCzpPhTRZ (133)**, **BzCzpPhTRZ–2 (135)** and **BzCzpPhTRZ–3 (136)** in neat thin film and doped in host materials.

Compounds	CzpPhTRZ (131)[c]	CyCzpPhTRZ (132)	TfCzpPhTRZ (133)	BzCzpPhTRZ-2 (135)	BzCzpPhTRZ-3 (136)
			Neat film		
λ_{abs} [nm][a]	_[d]	366	366	368	364
λ_{PL} [nm][a]	476	456	448	460	456
PLQY[%][b]	40	32	36	7	3
			10 wt% doped in DPEPO		
λ_{abs} [nm][a]	_[d]	372	370	368	350
λ_{PL} [nm][a]	482[e]	452	448	454	443
PLQY[%][b]	69[e]	50	49	13	5

[a] measurements completed under vacuum. [b] measurements completed under nitrogen and values were determined using an integrating sphere. All thin films coated by spin coating technique on quartz substrate and all solution made by using CHCl₃ as solvent. [c] data from literature.[78] [d] not available. [e] prepared by vacuum deposition and values were determined using an integrating sphere (λ_{exc} = 360 nm).

In summary, the emitters based on **CzpPhTRZ (131)**, a chiral TADF molecule which possesses through-space characteristics at the donor function, were designed and synthesized. Through the introduction of electron-withdrawing groups such as cyano (CN), trifluoromethyl (CF₃) and benzoyl groups (Bz) to the donor moiety, the emission wavelengths of the derivatives are blue-shifted from 482 nm to 443–454 nm. Particularly, **CyCzpPhTRZ (132)** and **TfCzpPhTRZ (133)** exhibit deep blue emission with λ_{PL} of 452 nm and 448 nm, while the PLQYs can still reach up to 50% and 49%, respectively. The TADF properties and potential electroluminescence applications of the deep blue emitters, **CyCzpPhTRZ (132)** and **TfCzpPhTRZ (133),** are under investigation.

3.3.2 Dicarbazolophane-Based TADF Emitters

Through the introduction of electron-withdrawing groups to **Czp (130)**, blue emitters with colors from 443 nm to 454 nm were obtained (Chapter 3.3.1). Though the PLQYs reach up to 49%, the ΔE_{ST} values of these emitters (0.31–0.38 eV) are relatively large compared with some other TADF emitters in literature.[166-171] In order to realize a more effective RISC for TADF, it is necessary to further decrease the ΔE_{ST} of the emitters. As mentioned before, a larger donor unit can more effectively disperse the HOMO away from the LUMO and reduce the spatial overlap, resulting in a smaller ΔE_{ST}.[172] Therefore, the emitters based on dicarbazolophane (**DCCP, 148**), a donor unit derived from carbazole (**75**) and **Czp (130)**, are investigated in this chapter (Figure 68).

Figure 68. Chemical structures of carbazole (**75**), **Czp (130)** and **DCCP (148)**.

3.3.2.1 Molecular Design and DFT Calculation

The emitter **DCCPTRZ (149)** with **DCCP (148)** as donor and 2,4-bis(4-(*tert*-butyl)phenyl)-1,3,5-triazine as acceptor was initially designed to explore its TADF properties (Figure 69). As identified in chapter 3.1, two *tert*-butyl groups were introduced to the acceptors as they show beneficial electron-withdrawing strength, extend the length of the target emitter in order to get preferential orientation and maintain solubility.

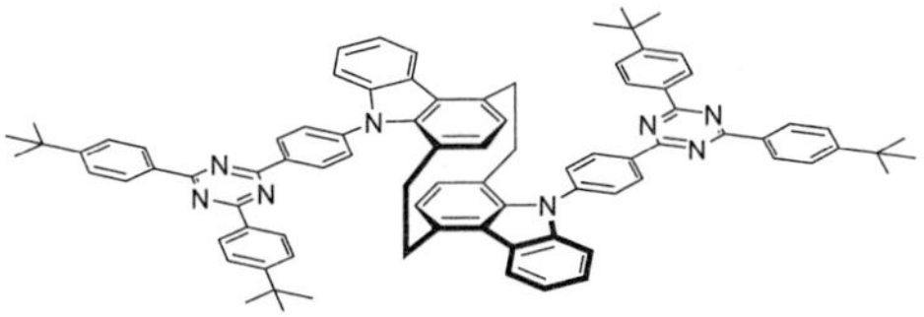

Figure 69. Chemical structure of **DCCPTRZ (149)**.

DFT calculations were performed employing the Gaussian 09 revision D.018 suite to investigate the frontier orbit distributions and energy states. First, the PBE0 functional with the standard Pople

6-31G (d,p) basis set was used to optimize the geometries of the molecular structures in the ground state in gas phase. Next, time-dependent DFT calculations were performed using the Tamm–Dancoff approximation (TDA) based on the optimized molecular structures. GaussView 5.0 software was employed to visualize the molecular orbitals. The frontier orbitals (HOMO/LUMO) and energy levels of **DCCPTRZ (149)** are shown in Figure 70 and Table 26.

The HOMO of **DCCPTRZ (149)** is mainly dispersed on the **DCCP (148)** donor unit with minor contributions on the phenyl bridges, while the LUMO is localized on the triazine parts with significant contributions on the phenyl bridges (Figure 70). HOMO and LUMO partly overlap on the phenyl bridges. Compared with **CzpPhTRZ (131)**, the ΔE_{ST} of **DCCPTRZ (149)** is decreased from 0.31 eV to 0.27 eV. Though the calculated S_1 and T_1 of **DCCPTRZ (149)** is slightly reduced, the oscillator strength (*f*) of **DCCPTRZ (149)** is significantly increased from 0.3420 for **CzpPhTRZ (131)** to 0.7959.

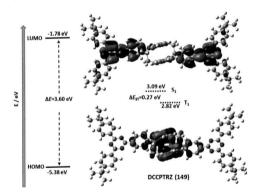

Figure 70. DFT calculations of **DCCPTRZ (149)**. Results were provided by Ettore Crovini.

Table 26. DFT calculations of **DCCPTRZ (149)** and **CzpPhTRZ (131)**.

Compounds	HOMO/LUMO [eV]	ΔE [eV]	*f*	S_1 [eV]	T_1 [eV]	ΔE_{ST} [eV]
DCCPTRZ (149)	−5.38/−1.78	3.60	0.7959	3.09	2.82	0.27
CzpPhTRZ (131)[a]	−5.54/−1.88	3.66	0.3420	3.27	2.96	0.31

[a] From Chapter 3.3.1.

3.3.2.2 Synthesis of DCCPTRZ

Based on the robust synthetic route established for **Czp (130)** in Chapter 3.3.1, **DCCP (148)** was synthesized *via* the identical two-step procedure (Scheme 33 and Scheme 34).

First, the secondary chloramine **152** was obtained *via* a Buchwald-Hartwig cross-coupling reaction of pseudo-*para* dibromide **150** with 2-chloroaniline **151** with a good yield of 76% (Scheme 33). A crystal structure of intermediate **152** was obtained, which confirmed its absolute configuration. From the structure, the distances between atoms of the two benzene rings WERE determined to be between 2.783 Å and 3.115 Å (Figure 71).

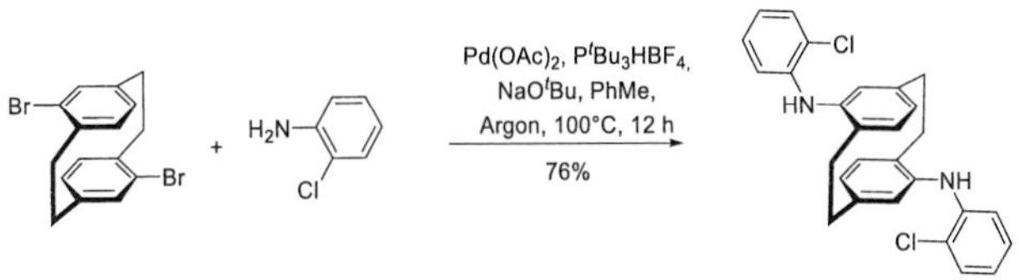

Scheme 33. Synthesis of **152** through Buchwald-Hartwig cross-coupling.

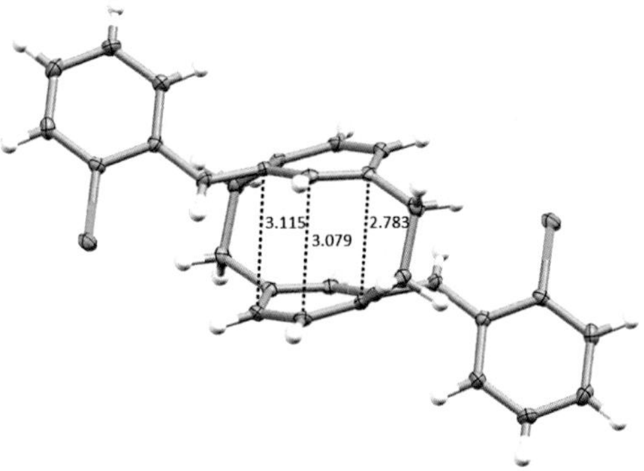

Figure 71. Molecular structure of **152** drawn at 50% probability level.

Then, through a two-fold Pd-catalyzed oxidative cyclization using the chlorine atoms as the directing groups, **DCCP (148)** was obtained in a yield of 49% (2.10 g scale, Scheme 34). The through-space dimer was fully characterized by NMR and HRMS. In addition, the absolute configuration of **DCCP (148)** was confirmed by single crystal analysis (Figure 74). The distances between the atoms of the two benzene rings were determined in the range of 2.761 Å to 3.056 Å. Compared with the conventional [2.2]Paracyclophane (129), the deck distance is reduced from 2.78 Å to 2.761 Å and 3.09 Å to 3.056 Å, suggesting that the two decks of the [2.2]cyclophane are closer to each other, facilitating a more effective through-space electronic communication within the donor unit.

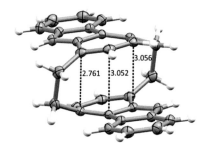

Scheme 34. Synthesis of **DCCP (148)** through a two-fold oxidative cyclization process.

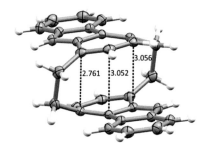

Figure 72. Molecular structure of **DCCP (148)** drawn at 50% probability level.

Lastly, the **DCCP (148)** donor and two equivalents of the fluorotriazine-based acceptor **98** were connected together *via* a nucleophilic aromatic substitution using tripotassium phosphate as base in DMSO at 120 °C to give **DCCPTRZ (149)** in good yield as a white solid with strong blue luminescence under UV excitation (Scheme 35).

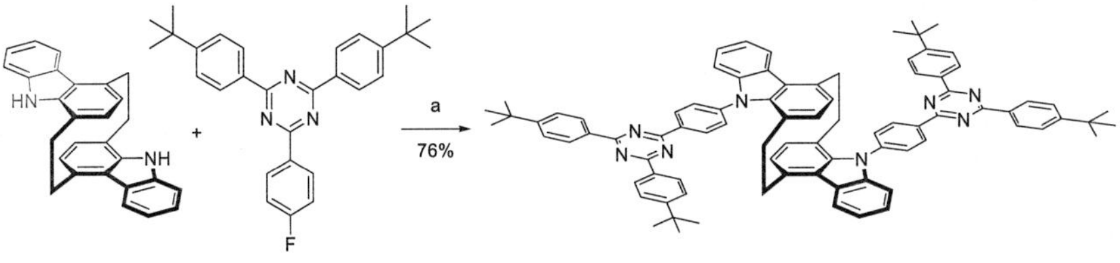

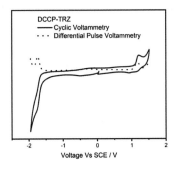

Scheme 35. Synthesis of **DCCPTRZ (149)**. a) K₃PO₄, Dry DMSO, Argon, 120 °C, 12 h.

3.3.2.3 Electrochemical and Photophysical Properties of DCCPTRZ

The electrochemical properties of **DCCPTRZ (149)** were examined by cyclic voltammetry (CV) and differential pulse voltammetry (DPV) (Figure 73), and the results are summarized in Table 27. The reduction process at −1.81 eV was assigned to the triazine-based acceptor part due to its electron-withdrawing property. The oxidation wave at 1.08 eV of this emitter is irreversible and is therefore assigned to the **DCCP (148)** donor unit due to the electrochemical instability of the carbazole radical cation. In addition, the optical gap of **DCCPTRZ (149)** was determined to 2.88 eV, showing potential for blue emission.

Figure 73. Cyclic Voltammetry (CV) and Differential Pulse Voltammetry (DPV) of **DCCPTRZ (149)** corrected against the Saturated Calomel Electrode (SCE).

Table 27. Electrochemical properties of **DCCPTRZ (149)**.

Compounds	Oxidation potential/[eV][a]	HOMO [eV]	Reduction potential/[eV][a]	LUMO [eV]	ΔE [eV]
DCCPTRZ (149)	1.08	−5.88	−1.81	−2.99	2.89

[a] From DPV.

Then, host material screenings were conducted to find the optimal host to obtain the best photoluminescent performances. The carbazole derived mCP, DPEPO and PMMA were tested because of their excellent charge transport abilities and high energy states. The doping concentration, which is also very crucial to the emission yield, was varied from 1 wt% to 15 wt%. As shown in Table 28. A trend of increased PLQYs was observed when the measurements were conducted under nitrogen atmosphere instead of air condition. The PLQYs also depend on the host materials and concentrations. For example, in mCP the quantum yield varied from 24.9% in 10 wt% to 42.8% in 4 wt%. For different host materials, the quantum yields could be increased from 7.0% to 37.4% by using mCP instead of DPEPO.

Table 28. Host optimization of spin-coated films for **DCCPTRZ (149)**.

Entry	Host material	Doping concentration	$\Phi_{PL}^{thin\ film}$ air/ %	$\Phi_{PL}^{thin\ film}$ nitrogen/ %
1		1 wt%	30.3	32.1
2		2 wt%	33.6	35.0
3		3 wt%	28.0	30.4
4	mCP	4 wt%	42.7	42.8
5		5 wt%	37.1	37.4
6		10 wt%	24.7	24.9
7		15 wt%	25.2	26.5
8		3 wt%	10.0	11.0
9	DPEPO	5 wt%	7.0	7.0
10		10 wt%	28.0	27.7
11		15 wt%	21.3	21.5
12	PMMA	10 wt%	19.5	20.4

Photoluminescence behaviors of **DCCPTRZ (149)** in neat film and in different host materials were measured (Figure 74). The photoluminescence emission maximum in neat film at 464 nm was blue shifted to 457 nm and 443 nm, when the emitter was doped in unpolar hosts such as PMMA and mCP, respectively. Also, in mCP the quantum yield could be increased to 42.8%, accompanied with deep blue emission (λ_{PL} = 443 nm). Time-resolved photoluminescence behavior of **DCCPTRZ (149)** in film was performed to further investigate its emission properties. The results are summarized in Table 29. As shown in Figure 74, a multiexponential fitting to the emission decay curve reveals a prompt decay component in the nanosecond range from 7.44 ns to 9.43 ns. However, no delayed component is observed for the emission, showing the absence of any dominant TADF emission in these tested films.

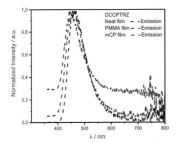 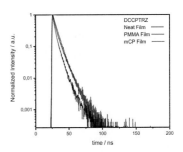

Figure 74. PL spectra (left) and photoluminescence decay curves (right) of **DCCPTRZ (149)**.

Table 29. Photophysical properties of **DCCPTRZ (149)** in neat thin film and doped in host materials.

Entry	host	$\lambda_{PL}^{\text{thin film}}$ / nm	$\Phi_{PL}^{\text{thin film}}$ air / %	$\Phi_{PL}^{\text{thin film}}$ nitrogen / %	τ_p / ns	τ_d / µs
1	neat	464	19.1	20.9	7.44	No delayed
2	PMMA	457	19.5	20.4	9.24	No delayed
3	mCP (4 wt%)	443	42.7	42.8	9.43	No delayed

3.3.2.4 Enlarged-DCCP Donors for TADF Emitters

In order to activate the TADF process of **DCCP**-based emitters, a further enlargement of the conjugation size of the donor part with the introduction of more electron-donating and conjugating units is an efficient strategy reported by Adachi *et al.* Therefore, two new molecules ***t*BuCzDCCPTRZ (154)** and **MeOCzDCCPTRZ (155)** were designed by adding carbazole derivatives to the **DCCP** core. For triazine-based acceptor units, more *tert*-butyl groups are necessary to be introduced to maintain solubility of these larger emitter molecules compared with **DCCPTRZ (149)**. **DCCPTRZ-2 (153)** was also designed in order to have a comparison in computations and experiments (Figure 75).

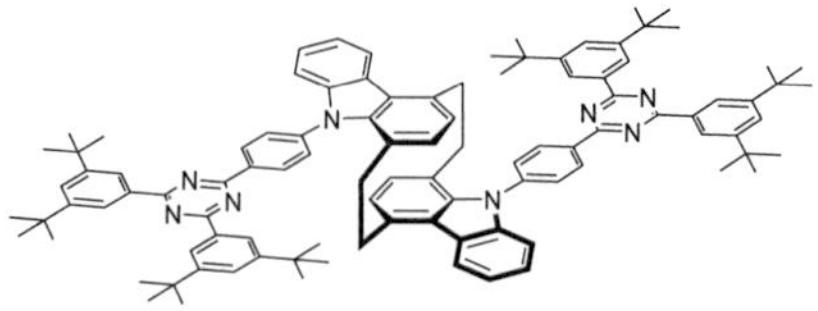

153
DCCPTRZ-2

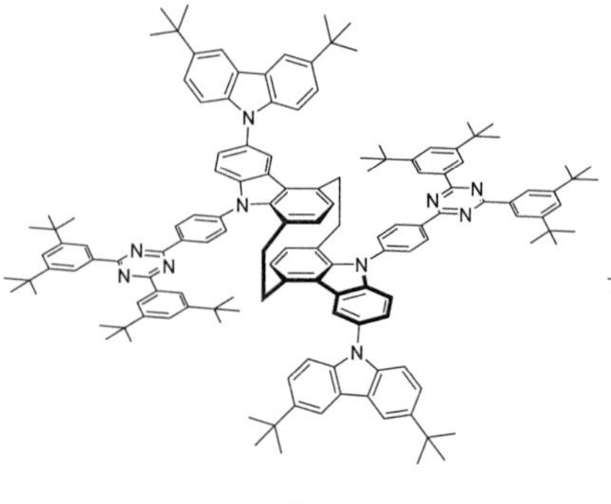

154
***t*BuCzDCCPTRZ**

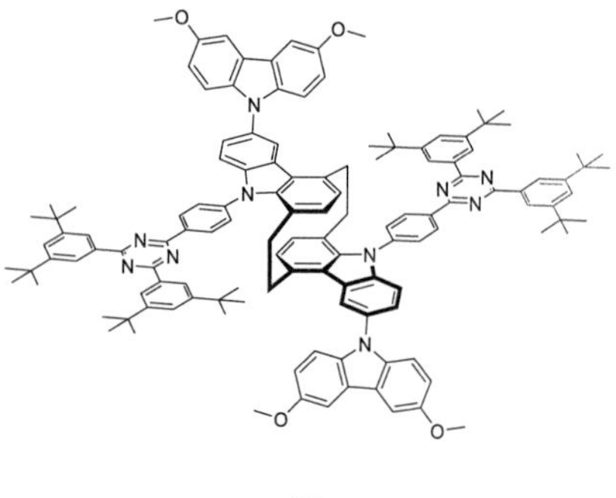

155
MeOCzDCCPTRZ

Figure 75. Chemical structures of **DCCPTRZ-2 (153), *t*BuCzDCCPTRZ (154)** and **MeOCzDCCPTRZ (155).**

DFT calculations were performed to investigate the frontier orbital distributions and energy states of these three emitters. As shown in Figure 76, the LUMO is localized on the triazine parts with significant contributions on the phenyl bridges. The HOMO of **DCCPTRZ–2 (153)** is mainly dispersed on the whole **DCCP** unit while the HOMO of ***t*BuCzDCCPTRZ (154)** and

MeOCzDCCPTRZ (155) is delocalized on the indolo part of **DCCP** and the appended methoxycarbazole-based units. From comparison of the ΔE_{ST} values, it is clear that the ΔE_{ST} drastically decreases through the use of larger donors. The ΔE_{ST} of **DCCPTRZ–2 (153)** is 0.28 eV while it is 0.16 eV for *t*BuCzDCCPTRZ (154). Furthermore, the ΔE_{ST} of **MeOCzDCCPTRZ (155)** was tuned to 0.11 eV due to the stronger electron-donating ability of 3,6-dimethoxy-9H-carbazole than 3,6-di-tert-butyl-9H-carbazole. Additionally, the HOMO of *t*BuCzDCCPTRZ (154) and **MeOCzDCCPTRZ (155)** is lifted from –5.39 eV (**DCCPTRZ–2, 153**) to –5.17 eV and –5.06 eV, respectively, induced by the addition of appended carbazole-based units.

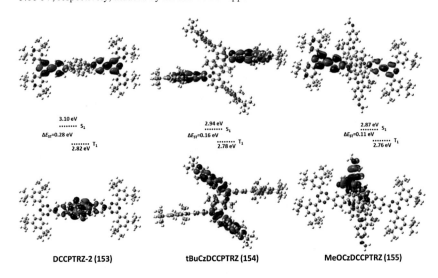

Figure 76. DFT excited state energies and electron density distributions of the HOMOs and LUMOs for **DCCPTRZ-2 (153)**, *t*BuCzDCCPTRZ (154) and **MeOCzDCCPTRZ (155)**.

Table 30. DFT calculations of **DCCPTRZ-2 (153)**, *t*BuCzDCCPTRZ (154) and **MeOCzDCCPTRZ (155)**.

Compounds	HOMO/LUMO [eV]	ΔE [eV]	f	S_1 [eV]	T_1 [eV]	ΔE_{ST} [eV]
DCCPTRZ–2 (153)	–5.39/–1.75	3.64	0.80	3.10	2.82	0.28
*t*BuCzDCCPTRZ (154)	–5.17/–1.82	3.35	0.30	2.94	2.78	0.16
MeOCzDCCPTRZ (155)	–5.06/–1.81	3.25	0.20	2.87	2.78	0.11

3.3.2.5 Synthesis of Enlarged-DCCP Based Emitters

First, according to the procedure for **DCCPTRZ (149)** (Scheme 35), **DCCPTRZ-2** (153) was synthesized *via* a nucleophilic substitution reaction between **DCCP (148)** and triazine-based acceptor **94** with a yield of 62% (Scheme 36).

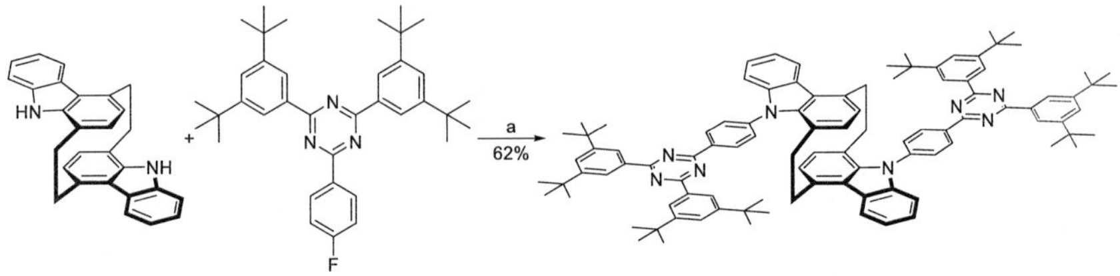

148
DCCP

94

153
DCCPTRZ-2

Scheme 36. Synthesis of **DCCPTRZ-2 (153)**. a) K_3PO_4, Dry DMSO, Argon, 120 °C, 12 h.

In order to synthesize ***t*BuCzDCCPTRZ (154)** and **MeOCzDCCPTRZ (155)**, bromination was conducted for **DCCP (148)** to introduce bromine atoms for the following Buchwald-Hartwig cross-coupling. Similar conditions as used in Scheme 23 (Chapter 3.3.1) was applied to prepare **156**. As expected, two bromine atoms were introduced to the **DCCP (148)**. The structure of **156** was fully characterized by NMR, MS, HRMS and IR.

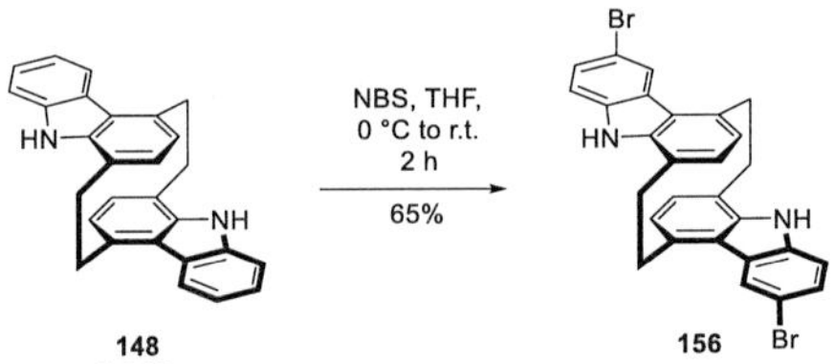

148
DCCP

156

Scheme 37. Bromination of **DCCP (148)** with NBS as the bromine source to generate dibromo-substituted compound **156**.

Furthermore, the absolute configuration of dibromo-substituted **156** was confirmed by single crystal analysis (Figure 77). The distances between the atoms of the two benzene rings are from 2.760 Å to 3.049 Å. Compared with the **DCCP (148)**, the distance is further slightly reduced from 2.761 Å to 2.760 Å and 3.052 Å to 3.049 Å, which means the two bent planes are closer to each other.

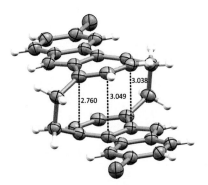

Figure 77. Molecular structure of **156** drawn at 50% probability level.

Then, **156** was reacted with **94** to generate **157** through nucleophilic substitution, and the two remaining bromine handles could be used for the introduction of additional carbazole-based units (Scheme 38).

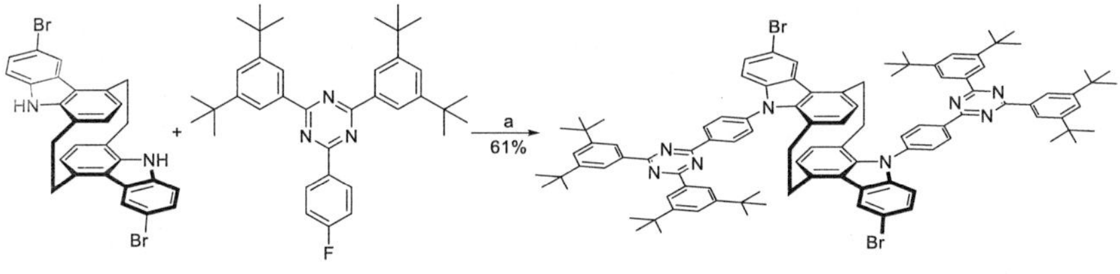

Scheme 38. Synthesis of **157**. a) K₃PO₄, Dry DMSO, Argon, 150 °C, 12 h.

Through the Buchwald-Hartwig cross-coupling process, two 3,6-di-*tert*-butyl-9*H*-carbazol groups were added to the **157**, generated the emitter *t***BuCzDCCPTRZ (154)** as a yellow solid with 75% yield (Scheme 39).

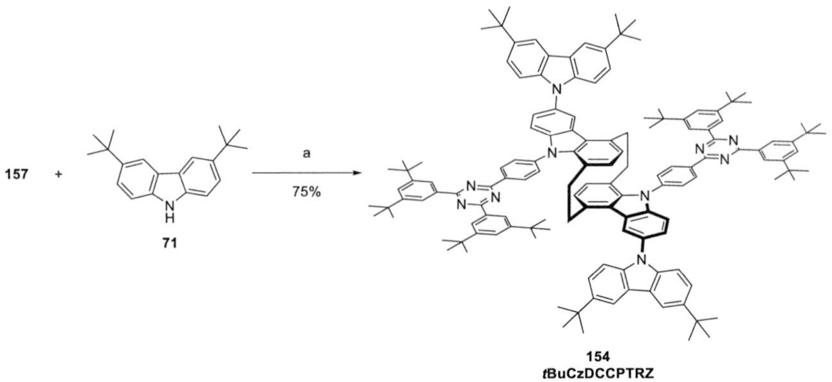

Scheme 39. Synthesis of *t*BuCzDCCPTRZ (**154**). a) Pd(OAc)$_2$, P^{t}Bu$_3$HBF$_4$, NaO^{t}Bu, PhMe, Argon, 100°C, 12 h.

Next, in order to prepare the **MeOCzDCCPTRZ (155)** derivative, the methoxy carbazole intermediate **159** was synthesized through copper(I) mediated methoxylation of the bromine substituents in a good yield of 72% as a yellow solid and strong blue luminescence was observed under UV excitation (Scheme 40).[173]

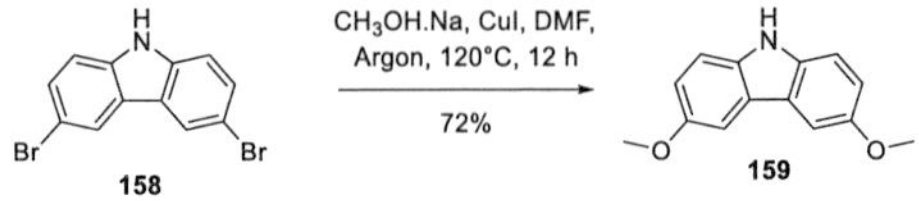

Scheme 40. Synthesis of dimethoxyl-substituted carbazole **159**.

Finally, through the Pd-catalyzed Buchwald-Hartwig cross-coupling, the emitter **MeOCzDCCPTRZ (155)** was synthesized with a yield of 74% as a yellow solid and strong blue luminescence was observed under 366 nm UV light (Scheme 41).

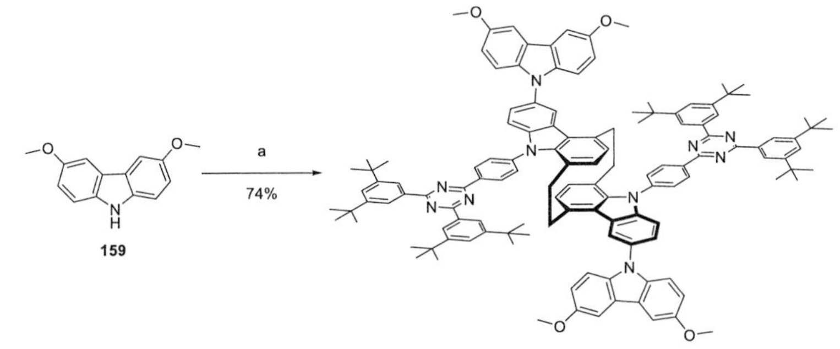

155
MeOCzDCCPTRZ

Scheme 41. Synthesis of **MeOCzDCCPTRZ (155)**. a) Pd(OAc)$_2$, P^{t}Bu$_3$HBF$_4$, NaO^{t}Bu, PhMe, Argon, 100°C, 12 h.

3.3.2.6 Electrochemical and Photophysical Properties

The electrochemical properties of these three emitters were examined by cyclic voltammetry (CV) and differential pulse voltammetry (DPV) (Figure 78), and the results are listed in Table 31. The reduction process at −1.62 eV to −1.79 eV is assigned to the triazine-based acceptor part due to its electron-withdrawing property. Interestingly, the oxidation waves of these three emitters show different characters. For **DCCPTRZ-2 (153)**, the oxidation wave at +1.19 V is irreversible, which is consistent with **DCCPTRZ (149)**. On the contrary, the oxidation waves of *t*BuCzDCCPTRZ **(154)** and **MeOCzDCCPTRZ (155)** at +1.09 V and +0.90 V, respectively, exhibit reversibility because the potential positions for dimerization on **DCCP** and the appended carbazole groups were substituted, resulting in electrochemical stability.[78, 174] In addition, the HOMO levels of *t*BuCzDCCPTRZ **(154)** and **MeOCzDCCPTRZ (155)** are lifted from −5.99 eV (**DCCPTRZ-2 (153)**) to −5.89 eV and −5.70 eV, respectively, due to the changed donor units, which is in line with theoretical calculations.

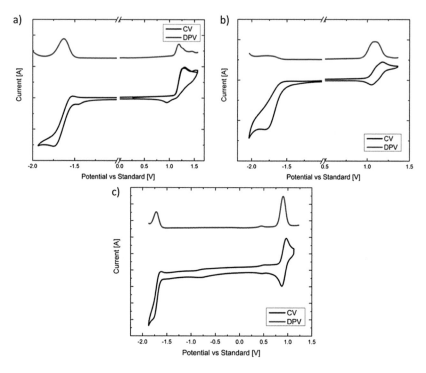

Figure 78. Cyclic Voltammetry (CV) and Differential Pulse Voltammetry (DPV) of **DCCPTRZ-2 (153, a),** *t*BuCzDCCPTRZ (154, b) and MeOCzDCCPTRZ (155, c) corrected against the Saturated Calomel Electrode (SCE). Results were provided by Stefan Diesing.

Table 31. Electrochemical properties of **DCCPTRZ-2 (153),** *t*BuCzDCCPTRZ (154) and MeOCzDCCPTRZ (155).

Compounds	Oxidation potential/[eV]	HOMO [eV]	Reduction potential/[eV]	LUMO [eV]	ΔE [eV]
DCCPTRZ-2 (153)	1.19	−5.99	−1.62	−3.18	2.81
*t*BuCzDCCPTRZ (154)	1.09	−5.89	−1.79	−3.01	2.88
MeOCzDCCPTRZ (155)	0.90	−5.70	−1.72	−3.08	2.62

Photoluminescence behaviors of **DCCPTRZ-2 (153)**, *t*BuCzDCCPTRZ **(154)** and MeOCzDCCPTRZ **(155)** in 10 wt% mCBP were measured (Figure 79). The absorption peaks of these three emitters between 350 to 400 nm can be ascribed to local transitions in the **DCCP** donor and triazine-based acceptor moieties. **DCCPTRZ-2 (153)** shows deep blue emission with the photoluminescence emission maximum at 452 nm and CIE coordinates of (0.16, 0.16). Though the emission of 463 nm with CIE coordinates at (0.17, 0.20) for *t*BuCzDCCPTRZ **(154)** and 478 nm with CIE coordinates at (0.20, 0.32) for **MeOCzDCCPTRZ (155)** was slightly red-shifted due to the introduction of appended carbazole-based units, these emitters are still very promising for blue OLEDs. Furthermore, *t*BuCzDCCPTRZ **(154)** exhibits a higher PLQY of 54% than **DCCPTRZ-2 (153)** of 50% and **MeOCzDCCPTRZ (155)** of 51%.

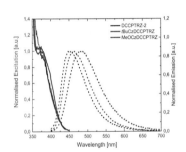 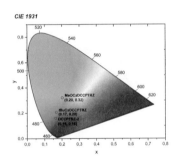

Figure 79. Left: UV-Vis and PL spectra of **DCCPTRZ-2 (153)**, *t*BuCzDCCPTRZ **(154)** and MeOCzDCCPTRZ **(155)** in 10 wt% mCBP. Right: CIE 1931 coordinates of the respective emitters. Results were provided by Stefan Diesing.

Time-resolved photoluminescence behaviors in 10 wt% mCBP were studied to reveal the emission properties of these three emitters. For *t*BuCzDCCPTRZ **(154)** and **MeOCzDCCPTRZ (155)**, the experimental results demonstrate that they are TADF emitters with delayed lifetimes of 43 μs and 35 μs (Figure 80 and Table 32), respectively, which is consistent with theoretical calculation of minute ΔE_{ST}. Unfortunately, no delayed emission was observed in 10 wt% mCBP using **DCCPTRZ-2 (153)** as dopant as only prompt fluorescence with a lifetime of 8.2 ns was found.

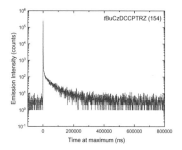 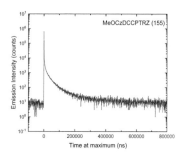

Figure 80. Photoluminescence decay curves of *t*BuCzDCCPTRZ **(154)** and MeOCzDCCPTRZ **(155)** in 10 wt% mCBP film. Results were provided by Stefan Diesing.

Table 32. Photophysical properties of **DCCPTRZ-2 (153)**, *t*BuCzDCCPTRZ **(154)** and MeOCzDCCPTRZ **(155)** in 10 wt% mCBP film.

Compounds	λ_{abs} [nm]	λ_{PL} [nm]	CIEy	PLQY[%]	τ_p / ns	τ_d / μs
DCCPTRZ-2 (153)	379	452	(0.16, 0.16)	50	8.2	No delayed
tBuCzDCCPTRZ (154)	381	463	(0.17, 0.20)	54	9.0	43
MeOCzDCCPTRZ (155)	374	478	(0.20, 0.32)	51	12.4	35

In summary, TADF emitters based on the novel through-space dimer **DCCP (148)** were designed and synthesized. Through the introduction of additional carbazole-based unit to the parent **DDCP(148)** donor to delocalize HOMO distributions, two blue TADF emitters, *t*BuCzDCCPTRZ **(154)** with CIE coordinates of (0.17, 0.20) and MeOCzDCCPTRZ **(155)** with CIE coordinates of (0.20, 0.32) could be accessed. Especially, the pure blue emitter *t*BuCzDCCPTRZ **(154)** exhibited λ_{PL} of 463 nm, a PLQY of 54% and a delayed lifetime of 43 μs, showing high potential for blue OLEDs. The electroluminescence performances of these two promising TADF emitters are under investigation.

4 Summary and Outlook

4.1 N-Heterocyclic Donors for Boron-Containing TADF Emitters

In the first part, six luminophores bearing the new OBO-fused benzo[fg]tetracene core as an electron-acceptor were investigated (Figure 81). DFT computations predicted that these emitters possess small S_1–T_1 energy gaps (ΔE_{ST}) in the range of 0.04 eV to 0.18 eV, accompanied by shallow LUMO levels.

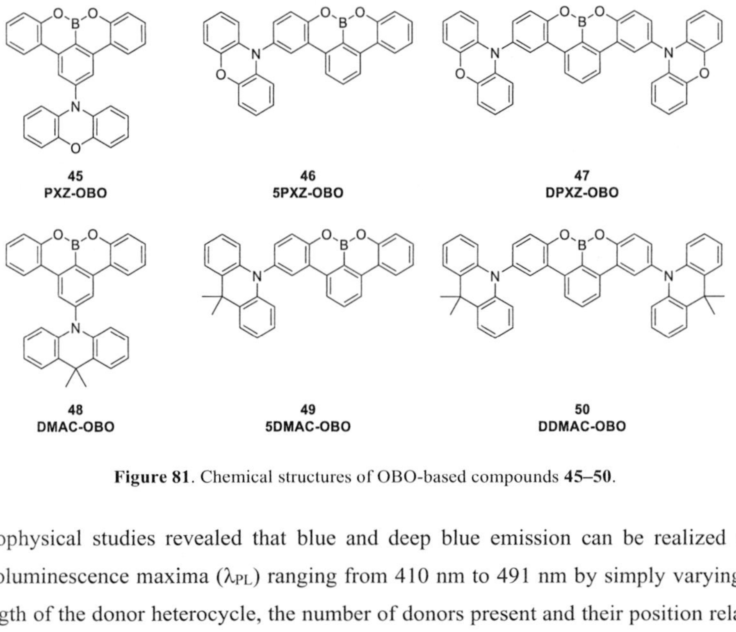

45
PXZ-OBO

46
5PXZ-OBO

47
DPXZ-OBO

48
DMAC-OBO

49
5DMAC-OBO

50
DDMAC-OBO

Figure 81. Chemical structures of OBO-based compounds **45–50**.

Photophysical studies revealed that blue and deep blue emission can be realized with photoluminescence maxima (λ_{PL}) ranging from 410 nm to 491 nm by simply varying the strength of the donor heterocycle, the number of donors present and their position relative to the acceptor (Figure 82). In doped films, the photoluminescence quantum yields (PLQYs) of these blue emitters reached up to 46% (**PXZ-OBO, 45**), which demonstrates their potential to be used as thermally activated delayed fluorescence (TADF) emitters for blue organic light-emitting diodes (OLEDs).

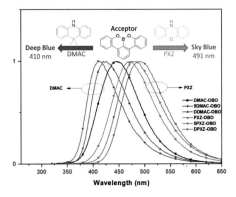

Figure 82. PL spectra of OBO-based emitters **45–50** in 10 wt% PMMA.

In addition, another boron-based unit was also used as acceptor to create TADF emitters. Through modifying a TADF emitter with introducing one more carbazole unit, the PLQY of **CzCzoB (68)** reaches up to 93% with a very short delayed lifetime of 3.1 μs, which is one of the shortest lifetime of triarylboron-derived TADF emitters (Figure 83). For the other emitter bearing an indolocarbazole as central donor, sky-blue emission with PLQY of 67% and delayed lifetime of 4.0 μs can be realized. The excellent photophysical properties of these two emitters suggest that they are good candidates for efficient TADF OLEDs. Further electroluminescence studies are currently under investigation.

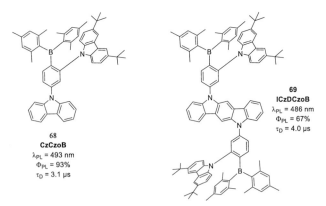

Figure 83. Chemical structures and photoluminescent properties of **CzCzoB (68)** and **ICzDCzoB (69)**.

4.1.1 Outlook for N-Heterocyclic Donors for Boron-Containing TADF Emitters

Though the OBO-based molecules show potential to be used as blue and deep blue TADF emitters, the low photoluminescence quantum yields and too long delayed lifetimes limit their applications for OLEDs. Therefore, the design and synthesis of novel boron-fused polycyclic units as acceptors for TADF emitters are meaningful. Some potential candidates (**160–163**) are exhibited in Figure 84.[116, 175]

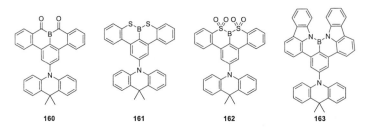

Figure 84. New designed TADF emitters based on boron-fused polycyclic acceptors.

4.2 Indolocarbazole-Based Donors for TADF Emitters

Secondly, a series of TADF emitters based on indolocarbazole was evaluated. The introduction of *tert*-butyl group to the triazine acceptor was proven to be a key factor to enhance the horizontal orientation in films, resulting in excellent TADF emitter for highly efficient OLEDs. Furthermore, it is confirmed that dimerization strategy was an effective method to tune the performance of TADF emitters. Additionally, the influence of steric hindrance between donor and acceptor and electron-withdrawing strength of acceptors to TADF emission were also explored in this part.

In the first place, a new linear acceptor-donor-acceptor molecule **ICzTRZ (85)** based on indolocarbazole and di-phenyltriazine unit were designed and synthesized (Figure 85). The single crystal analysis *via* X-Ray diffraction confirmed its linear structure. It showed nearly-completely horizontal orientation with the fraction of horizontal dipole Θ of 94%, leading to an OLED with EQE of 22.1% and sky-blue emission of 483 nm. Orientation studies demonstrated that the *tert*-butyl groups on the triazine-based acceptor played a key role to enhance the horizontal orientation.

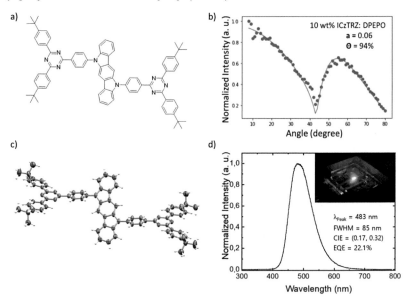

Figure 85. a) Chemical structure of **ICzTRZ (85)**. b) Orientation study in DPEPO. c) Molecular structure of **ICzTRZ (85)** drawn at 50% probability level. d) Device performances based on **ICzTRZ (85)**.

Then, considering the shared structure skeleton of **ICz (76)** and carbazole, the dimerization of **ICzTRZ (85)** and the TADF properties of the dimer were explored. Through the modification of a classic method for dimerization of carbazole derivatives with the increase of temperature, the dimer **DICzTRZ (104)** was successfully synthesized and photophysical results confirmed its TADF properties (Figure 86).

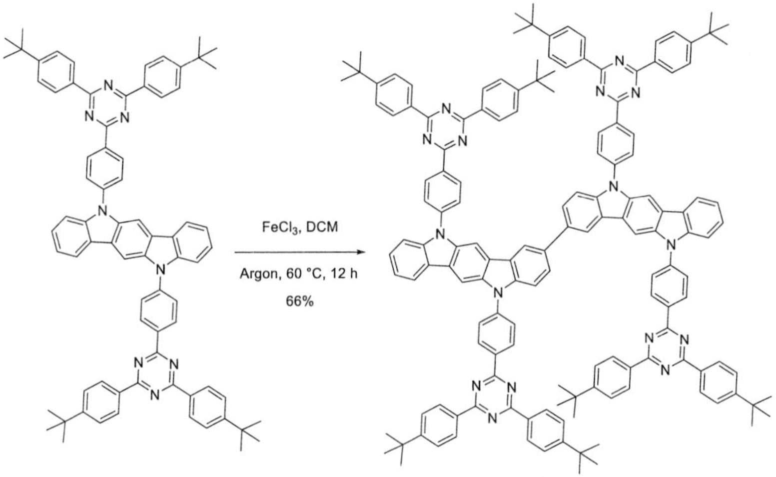

85
ICzTRZ

λ_{PL} = 472 nm (in 3 wt% mCP)
Φ_{PL} = 57.4 %
τ_D = 113.6 μs

FeCl$_3$, DCM
⟶
Argon, 60 °C, 12 h
66%

104
DICzTRZ

λ_{PL} = 486 nm (in 3 wt% mCP)
Φ_{PL} = 61.2%
τ_D = 237.9 μs

Figure 86. Dimerization of **ICzTRZ (85)** to generate **DICzTRZ (104)** and photophysical properties of them.

Lastly, in order to decrease the ΔE_{ST} of indolocarbazole-based emitters, the methyl groups were introduced to the phenyl bridges. The synthesis of the building block **125** was *via* a two-step one-pot sequential Buchwald-Hartwig amination/cyclization process with a high yield of 74%. The absolute structure of **125** shows a twist structure between phenyl unit and indolocarbazole with a dihedral angle of 84.52° (Figure 87).

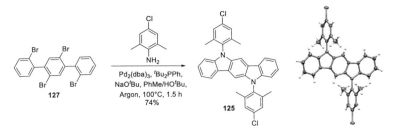

Figure 87. The synthesis of **125** *via* a two-step one-pot sequential Buchwald-Hartwig amination/cyclization process. The molecular structure of **125** drawn at 50% probability level.

Based on this building block, the molecules possessed twist structure between phenyl bridges and indolocarbazole were accessed (Figure 88). Experimental results indicated that photophysics of these molecules for TADF could be tuned. By varying the electron withdrawing strength of the acceptor moieties, **ICz-Me-PmTf (111)** realized TADF emission with delayed lifetime of 346 μs while no TADF emission was observed from **ICz-Me-PhMe (108)** and **ICz-Me-PmMe (110)** also showed low potential for TADF emission due to the weak electron-withdrawing of acceptor unit.

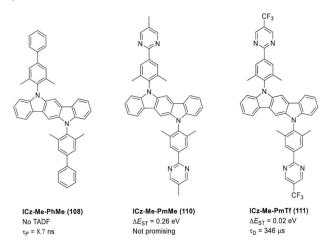

ICz-Me-PhMe (108)
No TADF
τ_P = 8.7 ns

ICz-Me-PmMe (110)
ΔE_{ST} = 0.26 eV
Not promising

ICz-Me-PmTf (111)
ΔE_{ST} = 0.02 eV
τ_D = 346 μs

Figure 88. Chemical structures and photophysical properties of **ICz-Me-PhMe (108)**, **ICz-Me-PmMe (110)** and **ICz-Me-PmTf (111)**.

4.2.1 Outlook for Indolocarbazole-Based Donors for TADF Emitters

Though **ICzTRZ (85)** showed excellent device performances enhanced by its nearly-completely horizontal orientation, the PLQY and delayed lifetime were not very satisfying. In order to improve these properties as well as maintain the linear structure, pyridyl-based unit can be used as linking bridge instead of phenyl unit to construct intramolecular hydrogen bonding channels to improve TADF properties as reported in literature.[176-180] Therefore, **164**, **165** and **166** are worth investigating further. For **166**, the methyl groups on pyridyl units can tune the dihedral angle between donor and acceptor (Figure 89).

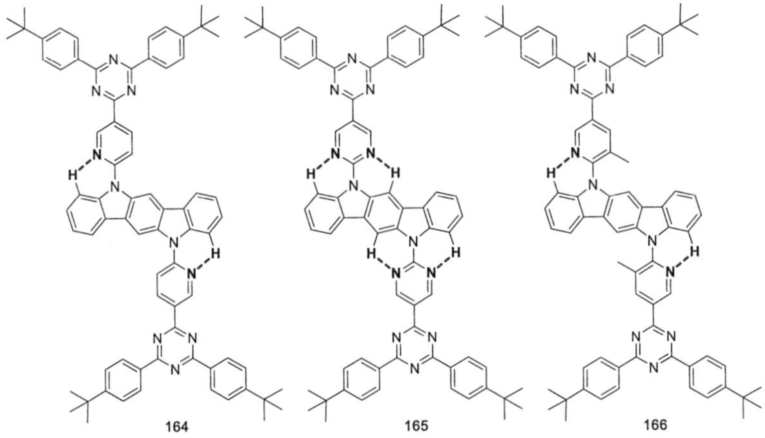

Figure 89. Molecular design based on the construction of intramolecular hydrogen bonding channels.

4.3 [2,2]Paracyclophane-Based Donors for TADF Emitters

The emitters based on **CyCzpPhTRZ (132)**, a chiral TADF molecule which possesses through-space interactions and additional steric bulk at the donor unit, were designed and synthesized. Through the introduction of electron-withdrawing groups such as cyano (CN), trifluoromethyl (CF₃) and benzoyl (Bz) groups to the donor moiety, the emission wavelengths were blue-shifted from 482 nm to 443–454 nm. Particularly, **CyCzpPhTRZ (132)** and **TfCzpPhTRZ (133)** exhibit deep blue emission with λ_{PL} of 452 nm and 448 nm while the PLQY remains at 50% and 49%, respectively (Figure 90).

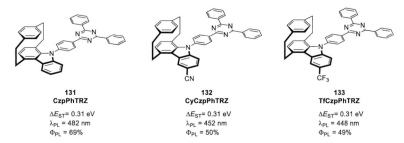

131
CzpPhTRZ

ΔE_{ST}= 0.31 eV
λ_{PL} = 482 nm
Φ_{PL} = 69%

132
CyCzpPhTRZ

ΔE_{ST}= 0.31 eV
λ_{PL} = 452 nm
Φ_{PL} = 50%

133
TfCzpPhTRZ

ΔE_{ST}= 0.31 eV
λ_{PL} = 448 nm
Φ_{PL} = 49%

Figure 90. Chemical structures and photoluminescent properties of **CyCzpPhTRZ (131)**, **CyCzpPhTRZ (132)** and **TfCzpPhTRZ (133)**.

Furthermore, in order to realize a more effective RISC for highly efficient TADF, the emitters based on the centrosymmetric through-space dimer **DCCP (148)**, a donor unit derived from carbazole (**73**) and **Czp (130)**, were investigated (Figure 91).

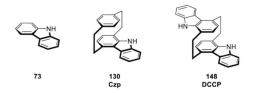

73

130
Czp

148
DCCP

Figure 91. Structures of carbazole (**73**), **Czp (130)** and **DCCP (148)**.

Through the introduction of appended carbazole-based units to the **DCCP (148)** donor to delocalize HOMO distributions, two blue TADF emitters, ***t*BuCzDCCPTRZ (154)** with CIE coordinates of (0.17, 0.20) and **MeOCzDCCPTRZ (155)** with CIE coordinates of (0.20, 0.32) were accessed. In particular, the pure blue emitter **tBuCzDCCPTRZ (154)** exhibits λ_{PL} of 463 nm, a PLQY of 54% and a delayed lifetime of 43 µs, showing high potential for blue OLEDs (Figure 92).

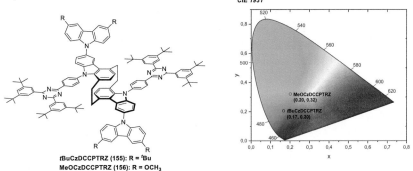

tBuCzDCCPTRZ (155): R = ᵗBu
MeOCzDCCPTRZ (156): R = OCH₃

Figure 92. Left: Chemical structures of **tBuCzDCCPTRZ (154)** and **MeOCzDCCPTRZ (155)**. Right: CIE 1931 coordinates of the respective emitters.

4.3.1 Outlook for [2,2]Paracyclophane-Based Donors for TADF Emitters

Due to the planar chirality of **Czp**-based molecules, they are promising candidates for circularly polarized luminescence (CPL).[78] Therefore, a method to synthesize and separate enantiopure donor units for TADF emitters is worth developing (Figure 93). Moreover, the investigation of the CPL properties of generated emitters for OLED applications is also meaningful and valuable.[181-185]

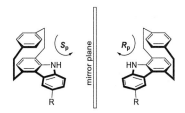

Figure 93. Enantiomers of **Czp**-based donors.

Furthermore, as mentioned before, the ΔE_{ST} of these **Czp**- and **DCCP**-based emitters are relatively large compared with some other TADF emitters in literature due to the small central five-membered N-heterocycle allowing a better conjugation due to lower sterical demand. Therefore, larger six-membered N-heterocycles can be used as donor units to further twist the donor and acceptor units while permitting through-space conjugation for other heterocyclic systems, yielding smaller ΔE_{ST}, accessing novel and promising TADF emitters as shown in Figure 94.

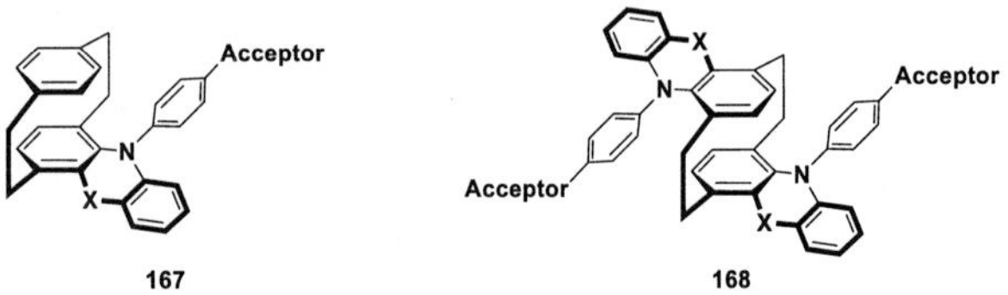

Figure 94. Six-membered N-heterocyclic donors for TADF emitters **167** and **168** (X = O, S, etc.).[186-187]

5 Experimental Section

5.1 General Information

5.1.1 Nomenclature of [2.2]Paracyclophanes

The IUPAC nomenclature for cyclophanes in general is rather confusing. Therefore Vögtle *et al.* developed a specific cyclophane nomenclature, which is based on a core-substituent ranking.[146] This is exemplified in Figure 95 for the [2.2]paracyclophane. [2.2]Paracylophane belongs to the D_{2h} point group, which is broken by the first substituent, resulting in two planar chiral enantiomers. For the determination of their absolute configuration, the arene bearing the substituent is set as the chiral phane, and the first atom of the cyclophane outside the phane and closest to the chirality center is set as the "pilot atom". If both arenes are substituted, then the substituent with higher priority according to the CAHN-INGOLD-PRELOG (CIP) nomenclature is preferred.[147] The stereo descriptor is determined by the sense of rotation viewed from the pilot atom.[148] An unambiguous numeration is required to describe the positions of the substituents correctly.

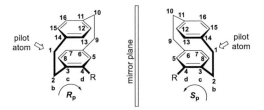

Figure 95. Nomenclature of mono-substituted [2.2]paracyclophane.

The first report of the [2]paracyclo[2](1,4)carbazolophane (**130**) by Bolm *et al.* presented the nomenclature, but no information on chirality was given.[151] The stereo descriptors follow the same standard as reported for the [2.2]paracyclophane. If viewed as a disubstituted [2.2]paracyclophane, then the nitrogen is assigned a higher priority than the carbon substituent, resulting in the respective sense of numbering as depicted in Figure 96. Based on this strategy, the nomenclature of the centrosymmetric through-space dimer donor (**148**) is also described in Figure 96.

Figure 96. Left: Nomenclature of the [2]paracyclo[2](1,4)carbazolophane (short: carbazolophane, **Czp**). Right: Nomenclature of the *trans-anti*-[2.2]-1,4-1',4'-dicarbazolocyclophane (dicarbazolocyclophane, **DCCP**).

5.1.2 Materials and Methods

Nuclear Magnetic Resonance Spectroscopy (NMR)

The NMR spectra of the compounds described herein were recorded on a Bruker Avance 300 NMR instrument at 300 MHz for ^{1}H NMR and 75 MHz for ^{13}C NMR, a Bruker Avance 400 NMR instrument at 400 MHz for ^{1}H NMR, 101 MHz for ^{13}C NMR and 376 MHz for ^{19}F NMR or a Bruker Avance 500 NMR instrument at 500 MHz for ^{1}H NMR, 125 MHz for ^{13}C NMR and 160 MHz for ^{11}B NMR.

The NMR spectra were recorded at room temperature in deuterated solvents acquired from Eurisotop. The chemical shift δ is displayed in parts per million [ppm] and the references used were the ^{1}H and ^{13}C peaks of the solvents themselves: d_1-chloroform (CDCl$_3$): 7.26 ppm for ^{1}H and 77.16 ppm for ^{13}C, d_6-dimethyl sulfoxide (DMSO-d_6): 2.50 ppm for ^{1}H and 39.4 ppm for ^{13}C and Tetrahydrofuran-d_8 (THF-d_8): 1.72 and 3.58 ppm for ^{1}H and 67.2 and 25.3 ppm for ^{13}C.

For the characterization of centrosymmetric signals, the signal's median point was chosen, for multiplets the signal range. The following abbreviations were used to describe the proton splitting pattern: s = singlet, d = doublet, t = triplet, m = multiplet, dd = doublet of doublet, ddd = doublet of doublet of doublet, dt = doublet of triplet. Absolute values of the coupling constants "J" are given in Hertz [Hz] in absolute value and decreasing order. Signals of the ^{13}C spectrum were assigned with the help of distortionless enhancement by polarization transfer spectra DEPT90 and DEPT135 and were specified in the following way: DEPT: + = primary or tertiary carbon atoms (positive DEPT signal), – = secondary carbon atoms (negative DEPT signal), C_q = quaternary carbon atoms (no DEPT signal).

Infrared Spectroscopy (IR)

The infrared spectra were recorded with a Bruker, IFS 88 instrument. Solids were measured by attenuated total reflection (ATR) method. The positions of the respective transmittance bands are given in wave numbers $\bar{\upsilon}$ [cm^{-1}] and was measured in the range from 3600 cm^{-1} to 500 cm^{-1}.

Characterization of the transmittance bands was done in sequence of transmission strength T with following abbreviations: vs (very strong, 0–9% T), s (strong, 10–39% T), m (medium, 40–69% T), w (weak, 70–89% T), vw (very weak, 90–100% T) and br (broad).

Mass Spectrometry (MS)

Electron ionization (EI) and fast atom bombardment (FAB) experiments were conducted using a Finnigan, MAT 90 (70 eV) instrument, with 3-nitrobenzyl alcohol (3-NBA) as matrix and reference for high resolution. For the interpretation of the spectra, molecular peaks [M]$^+$, peaks of protonated molecules [M+H]$^+$ and characteristic fragment peaks are indicated with their mass-to-charge ratio (m/z) and in case of EI their intensity in percent, relative to the base peak (100%) is given. In case of high-resolution measurements, the tolerated error is 0.0005 m/z.

APCI and ESI experiments were recorded on a Q-Exactive (Orbitrap) mass spectrometer (Thermo Fisher Scientific, San Jose, CA, USA) equipped with a HESI II probe to record high resolution. The tolerated error is 5 ppm of the molecular mass. Again, the spectra were interpreted by molecular peaks [M]$^+$, peaks of protonated molecules [M+H]$^+$ and characteristic fragment peaks and indicated with their mass-to-charge ratio (m/z).

Elemental Analysis (EA)

Elemental analysis was done on an Elementar vario MICRO instrument. The weight scale used was a Sartorius M2P. Calculated (calc.) and found percentage by mass values for carbon, hydrogen, nitrogen and sulfur are indicated in fractions of 100%.

Thin Layer Chromatography (TLC)

For the analytical thin layer chromatography, TLC silica plates coated with fluorescence indicator, from Merck (silica gel 60 F254, thickness 0.2 mm) were used. UV-active compounds were detected at 254 nm and 366 nm excitation wavelength with a Heraeus UV-lamp, model Fluotest.

Solvents and Chemicals

Solvents of p.a. quality (per analysis) were commercially acquired from Sigma Aldrich, Carl Roth or Acros Fisher Scientific and, unless otherwise stated, used without further purification. Dry solvents were either purchased from Carl Roth, Acros or Sigma Aldrich (< 50 ppm H_2O over molecular sieves). All reagents were commercially acquired from abcr, Acros, Alfa Aesar, Sigma Aldrich, TCI, Chempur, Carbolution or Synchemie, or were available in the group. Unless otherwise stated, all chemicals were used without further purification.

Experimental Procedure

Air- and moisture-sensitive reactions were carried out under argon atmosphere in previously baked out glassware using standard Schlenk techniques. Solid compounds were ground using a mortar and pestle before use, liquid reagents and solvents were injected with plastic syringes and stainless-steel cannula of different sizes, unless otherwise specified.

Reactions at low temperature were cooled using shallow vacuum flasks produced by Isotherm, Karlsruhe, filled with a water/ice mixture for 0 °C, water/ice/sodium chloride for –20 °C or isopropanol/dry ice mixture for –78 °C. For reactions at high temperature, the reaction flask was equipped with a reflux condenser and connected to the argon line.

Solvents were evaporated under reduced pressure at 40 °C using a rotary evaporator. Unless otherwise stated, solutions of inorganic salts are saturated aqueous solutions.

Reaction Monitoring

The progress of the reaction in the liquid phase was monitored by TLC. UV active compounds were detected with a UV-lamp at 254 nm and 366 nm excitation wavelength. When required, vanillin solution, potassium permanganate solution or methanolic bromocresol green solution was used as TLC-stain, followed by heating. Additionally, APCI-MS (atmospheric pressure chemical ionization mass spectrometry) was recorded on an Advion expression CMS in positive ion mode with a single quadrupole mass analyzer. The observed molecule ion is interpreted as $[M+H]^+$.

Product Purification

Unless otherwise stated, the crude compounds were purified by column chromatography. For the stationary phase of the column, silica gel, produced by Merck (silica gel 60, 0.040 × 0.063 mm, 260 – 400 mesh ASTM), and sea sand by Riedel de-Haën (baked out and washed with hydrochloric acid) were used. Solvents used were commercially acquired in HPLC-grade and individually measured volumetrically before mixing.

5.2   Analytical Data of N-Heterocyclic Donors for Boron-Containing TADF Emitters

10-(3,5-Dichlorophenyl)-10N-phenoxazine (53)

 A 20 mL sealed vial was charged with 10H-phenoxazine (274 mg, 1.50 mmol, 1.00 equiv.), 1-bromo-3,5-dichlorobenzene (356 mg, 1.57 mmol, 1.05 equiv.), palladium(II) acetate (6.7 mg, 29.9 µmol, 2 mol%), 4,5-bis(diphenylphosphino)-9,9-dimethylxanthene (17.0 mg, 29.4 µmol, 2 mol%) and sodium *tert*-butoxide (216 mg, 2.25 mmol, 1.50 equiv.). It was evacuated and flushed with argon three times. Through the septum 8 mL toluene was added, then it was heated to 100 °C and stirred for 12 h. The reaction mixture was diluted with 30 mL of ethyl acetate and washed with brine (20 mL). The organic layer was dried over MgSO₄ and the solvent was removed under reduced pressure. The obtained crude product was purified *via* flash-chromatography on silica gel (cyclohexane = 1) to yield the product as a white solid (405 mg, 1.23 mmol, 83%).

R_f = 0.60 (cyclohexane = 1). – **^{1}H NMR** (400 MHz, CDCl₃) δ = 7.49 (t, J = 1.9 Hz, 1H), 7.30 (d, J = 1.9 Hz, 2H), 6.81–6.52 (m, 6H), 6.07–5.85 (m, 2H) ppm. – **^{13}C NMR** (101 MHz, CDCl₃) δ = 143.8 (C$_q$, 2C$_{Ar}$), 141.3 (C$_q$, C$_{Ar}$), 137.1 (C$_q$, 2C$_{Ar}$), 133.3 (C$_q$, 2C$_{Ar}$), 129.8 (+, 2C$_{Ar}$H), 129.1 (+, C$_{Ar}$H), 123.4 (+, 2C$_{Ar}$H), 122.1 (+, 2C$_{Ar}$H), 115.8 (+, 2C$_{Ar}$H), 113.3 (+, 2C$_{Ar}$H) ppm. – **IR** (ATR, ṽ) = 3070, 1575, 1562, 1487, 1465, 1425, 1402, 1329, 1293, 1272, 1207, 1184, 1122, 1094, 975, 919, 873, 800, 727, 711, 693, 669, 603, 446 cm⁻¹. – **MS** (EI), *m/z*: 328 [M+H]⁺, 327 [M]⁺. – **HRMS** (EI, C₁₈H₁₁O₁N₁^{35}Cl₂) calc.: 327.0218; found: 327.0218.

10-(3,5-Dichlorophenyl)-9,9-dimethyl-9,10-dihydroacridine (54)

 A 20 mL sealed vial was charged with 9,9-dimethyl-9,10-dihydroacridine (105 mg, 502 μmol, 1.00 equiv.), 1-bromo-3,5-dichlorobenzene (135 mg, 610 μmol, 1.21 equiv.), palladium(II) acetate (6.0 mg, 26.7 μmol, 5 mol%), 4,5-bis(diphenylphosphino)-9,9-dimethylxanthene (Xantphos, 14.5 mg, 25.1 μmol, 5 mol%) and sodium *tert*-butoxide (72.0 mg, 749 μmol, 1.49 equiv.). It was evacuated and flushed with argon three times. Through the septum 8 mL toluene was added, then it was heated to 100 °C and stirred for 12 h. The reaction mixture was diluted with 30 mL of ethyl acetate and washed with brine (20 mL). The organic layer was dried over $MgSO_4$ and the solvent was removed under reduced pressure. The obtained crude product was purified *via* flash-chromatography on silica gel (cyclohexane = 1) to yield the product as a white solid (113 mg, 319 μmol, 64%).

R_f = 0.65 (cyclohexane/ethyl acetate = 50:1). – **^{1}H NMR** (400 MHz, CDCl$_3$) δ = 7.53 (t, J = 1.9 Hz, 1H), 7.47 (dd, J = 7.6 Hz, J = 1.7 Hz, 2H), 7.29 (d, J = 1.9 Hz, 2H), 7.06–6.94 (m, 4H), 6.28 (dd, J = 8.1 Hz, J = 1.4 Hz, 2H), 1.67 (s, 6H) ppm. – **^{13}C NMR** (101 MHz, CDCl$_3$) δ = 143.5 (C_q, C_{Ar}), 140.2 (C_q, 2C_{Ar}), 137.0 (C_q, 2C_{Ar}), 130.6 (C_q, 2C_{Ar}), 130.3 (+, 2C_{Ar}H), 128.9 (+, C_{Ar}H), 126.7 (+, 2C_{Ar}H), 125.6 (+, 2C_{Ar}H), 121.4 (+, 2C_{Ar}H), 114.2 (+, 2C_{Ar}H), 36.1 (C_q, CC_{Ar} (CH$_3$)$_2$), 31.3 (+, 2CH$_3$) ppm. – **IR** (ATR, ṽ) = 2966, 1589, 1571, 1561, 1479, 1460, 1455, 1429, 1412, 1388, 1327, 1281, 1268, 1232, 1221, 1194, 1105, 1099, 1089, 1045, 977, 929, 873, 858, 802, 745, 696, 684, 652, 612, 520, 472, 405 cm^{-1}. – **MS** (EI), *m/z*: 354 [M+H]$^+$, 353 [M]$^+$. – **HRMS** (EI, $C_{21}H_{17}N_1{}^{35}Cl_2$) calc.: 353.0738; found: 353.0736.

10-(2,2''-Dimethoxy-[1,1':3',1''-terphenyl]-5'-yl)-10N-phenoxazine (55)

A 20 mL sealed vial was charged with 10-(3,5-dichlorophenyl)phenoxazine (164 mg, 500 μmol, 1.00 equiv.), (2-methoxyphenyl)boronic acid (227 mg, 1.49 mmol, 2.99 equiv.), palladium(II) acetate (11.0 mg, 49.0 μmol, 10 mol%), 2-dicyclohexylphosphino-2′,6′-dimethoxybiphenyl (41.0 mg, 99.9 μmol, 20 mol%) and potassium phosphate tribasic (1.06 g, 5.00 mmol, 10.00 equiv.). It was evacuated and flushed with argon three times. Through the septum 10 mL of tetrahydrofuran and 3 mL of water were added, then it was heated to 80 °C and stirred for 12 h. The reaction mixture was diluted with 30 mL of ethyl acetate and washed with brine (20 mL). The organic layer was dried over MgSO$_4$ and the solvent was removed under reduced pressure. The obtained crude product was purified *via* flash-chromatography on silica gel (cyclohexane/dichloromethane = 15:1 to 5:1) to yield the product as a white solid (220 mg, 467 μmol, 93%).

R_f = 0.20 (cyclohexane/dichloromethane = 15:1). – **^{1}H NMR** (400 MHz, CDCl$_3$) δ = 7.80 (t, J = 1.6 Hz, 1H), 7.54 (d, J = 1.5 Hz, 2H), 7.44 (dd, J = 7.5 Hz, J = 1.7 Hz, 2H), 7.35 (td, J = 8.2 Hz, J = 1.7 Hz, 2H), 7.14–6.93 (m, 4H), 6.80–6.54 (m, 6H), 6.38–6.16 (m, 2H), 3.84 (s, 6H) ppm. – **^{13}C NMR** (101 MHz, CDCl$_3$) δ = 156.5 (+, 2C$_{Ar}$H), 144.2 (C$_q$, 2C$_{Ar}$), 140.9 (C$_q$, 2C$_{Ar}$), 137.7 (C$_q$, C$_{Ar}$), 134.6 (C$_q$, 2C$_{Ar}$), 130.8 (+, 2C$_{Ar}$H), 130.4 (+, 2C$_{Ar}$H), 130.2 (+, C$_{Ar}$H), 129.7 (C$_q$, 2C$_{Ar}$), 129.2 (+, 2C$_{Ar}$H), 123.3 (+, 2C$_{Ar}$H), 121.3 (+, 2C$_{Ar}$H), 121.0 (+, 2C$_{Ar}$H), 115.4 (+, 2C$_{Ar}$H), 113.7 (+, 2C$_{Ar}$H), 111.4 (+, 2C$_{Ar}$H), 55.7 (C$_q$, 2COCH$_3$) ppm. – **IR** (ATR, $\tilde{v}$) = 2839, 1587, 1497, 1482, 1462, 1455, 1431, 1414, 1324, 1288, 1272, 1262, 1242, 1200, 1181, 1163, 1119, 1047, 1026, 892, 742, 718, 670, 647, 615, 602, 567, 547, 520, 489, 475, 441, 424 cm^{-1}. – **MS** (EI), m/z: 472 [M+H]$^+$, 471 [M]$^+$. – **HRMS** (FAB, 3-NBA, C$_{32}$H$_{25}$O$_3$N$_1$) calc.: 471.1834; found: 471.1833.

10-(2,2''-Dimethoxy-[1,1':3',1''-terphenyl]-5'-yl)-9,9-dimethyl-9,10-dihydroacridine (56)

A 20 mL sealed vial was charged with 10-(3,5-dichlorophenyl)-9,9-dimethyl-9,10-dihydroacridine (88.0 mg, 248 µmol, 1.00 equiv.), (2-methoxyphenyl)boronic acid (114 mg, 750 µmol, 3.02 equiv.), palladium(II) acetate (5.6 mg, 24.9 µmol, 10 mol%), 2-dicyclohexylphosphino-2',6'-dimethoxybiphenyl (20.3 mg, 49.4 µmol, 20 mol%) and potassium phosphate tribasic (527 mg, 2.48 mmol, 10.00 equiv.). It was evacuated and flushed with argon three times. Through the septum 10 mL of tetrahydrofuran and 3 mL of water were added, then it was heated to 80 °C and stirred for 12 h. The reaction mixture was diluted with 30 mL of ethyl acetate and washed with brine (20 mL). The organic layer was dried over MgSO$_4$ and the solvent was removed under reduced pressure. The obtained crude product was purified *via* flash-chromatography on silica gel (cyclohexane/ethyl acetate = 20:1) to yield the product as a white solid (113 mg, 227 µmol, 91%).

R_f = 0.45 (cyclohexane/ethyl acetate = 20:1). – **^{1}H NMR** (400 MHz, CDCl$_3$) δ = 7.86 (t, J = 1.6 Hz, 1H), 7.52 (d, J = 1.6 Hz, 2H), 7.46 (td, J = 7.7 Hz, J = 7.1 Hz, J = 1.6 Hz, 4H), 7.34 (td, J = 8.2 Hz, J =1.7 Hz, 2H), 7.11–6.87 (m, 8H), 6.63 (dd, J = 8.2 Hz, J =1.1 Hz, 2H), 3.83 (s, 6H), 1.71 (s, 6H) ppm. – **^{13}C NMR** (101 MHz, CDCl$_3$) δ = 156.6 (C$_q$, 2C$_{Ar}$), 141.3 (C$_q$, 2C$_{Ar}$), 140.8 (C$_q$, 2C$_{Ar}$), 140.1 (C$_q$, 1C$_{Ar}$), 131.0 (+, 2C$_{Ar}$H), 130.9 (+, 2C$_{Ar}$H), 130.3 (+, C$_{Ar}$H), 130.2 (C$_q$, 2C$_{Ar}$), 130.0 (C$_q$, 2C$_{Ar}$), 129.1 (+, 2C$_{Ar}$H), 126.4 (+, 2C$_{Ar}$H), 125.0 (+, 2C$_{Ar}$H), 121.0 (+, 2C$_{Ar}$H), 120.5 (+, 2C$_{Ar}$H), 114.6 (+, 2C$_{Ar}$H), 111.4 (+, 2C$_{Ar}$H), 55.8 (+, 2CH$_3$O), 36.2 (C$_q$, CC$_{Ar}$(CH$_3$)$_2$), 31.0(+, 2CH$_3$) ppm. – **IR** (ATR, ṽ) = 2922, 1587, 1496, 1472, 1466, 1446, 1412, 1323, 1273, 1249, 1218, 1179, 1160, 1122, 1045, 1024, 904, 752, 714, 662, 630, 616, 443 cm^{-1}. – **MS** (EI), *m/z*: 498 [M+H]$^+$, 497 [M]$^+$. – **HRMS** (FAB, 3-NBA, C$_{35}$H$_{31}$O$_2$N$_1$) calc.: 497.2355; found: 497.2355.

10-(3-Chloro-4-methoxyphenyl)-10*N*-phenoxazine (57)

 A 20 mL sealed vial was charged with 10*H*-phenoxazine (137 mg, 750 μmol, 1.00 equiv.), 4-bromo-2-chloro-1-methoxybenzene (174 mg, 787 μmol, 1.05 equiv.), palladium(II) acetate (3.3 mg, 14.7 μmol, 2 mol%), 4,5-bis(diphenylphosphino)-9,9-dimethylxanthene (8.6 mg, 15.0 μmol, 2 mol%) and sodium *tert*-butoxide (108 mg, 1.12 mmol, 1.50 equiv.). It was evacuated and flushed with argon three times. Through the septum 8 mL toluene was added, then it was heated to 100 °C and stirred for 12 h. The reaction mixture was diluted with 30 mL of ethyl acetate and washed with brine (20 mL). The organic layer was dried over MgSO$_4$ and the solvent was removed under reduced pressure. The obtained crude product was purified *via* flash-chromatography on silica gel (cyclohexane/dichloromethane = 2:1 to 1:1) to yield the product as a white solid (217 mg, 670 μmol, 89%).

R_f = 0.35 (cyclohexane/dichloromethane = 2:1). – ^{1}H NMR (400 MHz, CDCl$_3$) δ = 7.38 (d, *J* = 2.5 Hz, 1H), 7.23 (dd, *J* = 8.6 Hz, *J* = 2.5 Hz, 1H), 7.12 (d, *J* = 8.7 Hz, 1H), 6.77–6.53 (m, 6H), 5.93 (dd, *J* = 7.6 Hz, *J* = 1.6 Hz, 2H), 3.98 (s, 3H) ppm. – ^{13}C NMR (101 MHz, CDCl$_3$) δ = 155.3 (C$_q$, C$_{Ar}$), 144.1 (C$_q$, 2C$_{Ar}$), 134.5 (C$_q$, 2C$_{Ar}$), 132.8 (+, C$_{Ar}$H), 132.0 (C$_q$, C$_{Ar}$), 130.6 (+, C$_{Ar}$H), 124.6 (C$_q$, C$_{Ar}$), 123.5 (+, 2C$_{Ar}$H), 121.7 (+, 2C$_{Ar}$H), 115.7 (+, 2C$_{Ar}$H), 113.9 (+, C$_{Ar}$H), 113.4 (+, 2C$_{Ar}$H), 56.6 (+, CH$_3$O) ppm. – IR (ATR, ṽ) = 3024, 2970, 1626, 1591, 1570, 1507, 1486, 1460, 1435, 1394, 1337, 1290, 1271, 1262, 1245, 1204, 1184, 1150, 1142, 1116, 1094, 1058, 1043, 1018, 958, 929, 885, 857, 837, 815, 744, 721, 707, 683, 633, 609, 589, 557, 537, 456, 416, 402 cm^{-1}. – MS (EI), *m/z*: 324 [M+H]$^+$, 323 [M]$^+$. – HRMS (EI, C$_{19}$H$_{14}$O$_2$N$_1$^{35}Cl$_1$) calc.: 323.0713; found: 323.0712.

10-(3-Chloro-4-methoxyphenyl)-9,9-dimethyl-9,10-dihydroacridine (58)

 A 20 mL sealed vial was charged with 9,9-dimethyl-10H-acridine (157 mg, 750 µmol, 1.00 equiv.), 4-bromo-2-chloro-1-methoxybenzene (174 mg, 787 µmol, 1.05 equiv.), palladium(II) acetate (3.4 mg, 15.1 µmol, 2 mol%), 4,5-bis(diphenylphosphino)-9,9-dimethylxanthene (8.7 mg, 15.0 µmol, 2 mol%) and sodium *tert*-butoxide (108 mg, 1.12 mmol, 1.50 equiv.). It was evacuated and flushed with argon three times. Through the septum 8 mL toluene was added, then it was heated to 100 °C and stirred for 12 h. The reaction mixture was diluted with 30 mL of ethyl acetate and washed with brine (20 mL). The organic layer was dried over MgSO$_4$ and the solvent was removed under reduced pressure. The obtained crude product was purified *via* flash-chromatography on silica gel (cyclohexane/ethyl acetate = 20:1) to yield the product as a white solid (260 mg, 743 µmol, 99%).

R$_f$ = 0.40 (cyclohexane/ethyl acetate = 20:1). – **¹H NMR** (400 MHz, CDCl$_3$) δ = 7.46 (dd, J = 7.6 Hz, J = 1.6 Hz, 2H), 7.39 (d, J = 2.4 Hz, 1H), 7.24 (dd, J = 8.6 Hz, J = 2.4 Hz, 1H), 7.16 (d, J = 8.6 Hz, 1H), 7.02–6.92 (m, 4H), 6.31 (dd, J = 8.1 Hz, J = 1.2 Hz, 2H), 4.02 (s, 3H), 1.69 (s, 6H) ppm. – **¹³C NMR** (101 MHz, CDCl$_3$) δ = 155.0 (C$_q$, C$_{Ar}$), 141.0 (C$_q$, 2C$_{Ar}$), 134.2 (C$_q$, C$_{Ar}$), 133.2 (+, C$_{Ar}$H), 130.8 (+, C$_{Ar}$H), 130.2 (C$_q$, 2C$_{Ar}$), 126.5 (+, 2C$_{Ar}$H), 125.4 (+, 2C$_{Ar}$H), 124.2 (C$_q$, C$_{Ar}$), 120.9 (+, 2C$_{Ar}$H), 114.1 (+, 2C$_{Ar}$H), 113.6 (+, C$_{Ar}$H), 56.5 (+, CH$_3$O), 36.1 (C$_q$, *C*C$_{Ar}$ (CH$_3$)$_2$), 31.4 (+, 2CH$_3$) ppm. – **IR** (ATR, ṽ) = 2962, 1589, 1494, 1475, 1438, 1395, 1387, 1326, 1285, 1275, 1262, 1239, 1183, 1164, 1145, 1129, 1111, 1089, 1058, 1047, 1021, 959, 929, 826, 800, 759, 749, 725, 698, 646, 612, 581, 569, 520, 429 cm^{-1}. – **MS** (EI), *m/z*: 350 [M+H]$^+$, 349 [M]$^+$. – **HRMS** (EI, C$_{22}$H$_{20}$O$_1$N$_1$^{35}Cl$_1$) calc.: 349.1233; found: 349.1231.

10-(2'',6-Dimethoxy-[1,1':3',1''-terphenyl]-3-yl)-10H-phenoxazine (59)

A 100 mL round flask was charged with 10-(3-chloro-4-methoxyphenyl)-10H-phenoxazine (1.36 g, 4.20 mmol, 1.00 equiv.), 2-(2'-methoxy-[1,1'-biphenyl]-3-yl)-4,4,5,5-tetramethyl-1,3,2-dioxaborolane (1.43 g, 4.61 mmol, 1.10 equiv.), palladium(II) acetate (47.0 mg, 209 μmol, 5 mol%), 2-dicyclohexylphosphino-2',6'-dimethoxybiphenyl (172 mg, 419 μmol, 10 mol%) and potassium phosphate tribasic (2.67 g, 12.6 mmol, 2.99 equiv.). It was evacuated and flushed with argon three times. Through the septum 50 mL toluene was added, then it was heated to 100 °C and stirred for 12 h. The reaction mixture was diluted with 50 mL of ethyl acetate and washed with brine (2 × 50 mL). The organic layer was dried over MgSO$_4$ and the solvent was removed under reduced pressure. The obtained crude product was purified *via* flash-chromatography on silica gel (cyclohexane/dichloromethane = 5:1 to 2:1) to yield the product as a white solid (1.46 g, 3.10 mmol, 74%).

R$_f$ = 0.35 (cyclohexane/dichloromethane = 5:1). – **^{1}H NMR** (400 MHz, CDCl$_3$) δ = 7.70 (t, J = 1.6 Hz, 1H), 7.58–7.50 (m, 2H), 7.45 (t, J = 7.6 Hz, 1H), 7.40–7.24 (m, 4H), 7.17 (d, J = 8.6 Hz, 1H), 7.08–6.96 (m, 2H), 6.75–6.52 (m, 6H), 6.04 (d, J = 6.2 Hz, 2H), 3.92 (s, 3H), 3.82 (s, 3H) ppm. – **^{13}C NMR** (101 MHz, CDCl$_3$) δ = 156.6 (C$_q$, C$_{Ar}$OCH$_3$), 156.4 (C$_q$, C$_{Ar}$OCH$_3$), 144.1 (C$_q$, C$_{Ar}$), 138.5 (C$_q$, 2C$_{Ar}$), 137.2 (C$_q$, C$_{Ar}$), 134.8 (C$_q$, C$_{Ar}$), 133.5 (C$_q$, C$_{Ar}$), 133.1 (+, C$_{Ar}$H), 131.6 (C$_q$, C$_{Ar}$), 131.1 (+, C$_{Ar}$H), 130.7 (C$_q$, 2C$_{Ar}$), 130.6 (+, 2C$_{Ar}$H), 128.9 (+, C$_{Ar}$H), 128.8 (+, C$_{Ar}$H), 128.1 (+, C$_{Ar}$H), 127.8 (+, C$_{Ar}$H), 123.4 (+, C$_{Ar}$H), 121.2 (+, C$_{Ar}$H), 120.9 (+, 2C$_{Ar}$H), 115.4 (+, C$_{Ar}$H), 113.4 (+, 2C$_{Ar}$H), 113.3 (+, C$_{Ar}$H), 111.3 (+, 2C$_{Ar}$H), 56.0 (+, CH$_3$O), 55.7 (+, CH$_3$O) ppm. – **IR** (ATR, ṽ) = 2925, 1592, 1482, 1460, 1439, 1398, 1384, 1329, 1290, 1269, 1244, 1221, 1200, 1179, 1162, 1132, 1118, 1094, 1051, 1043, 1027, 976, 955, 928, 914, 905, 891, 858, 799, 738, 704, 656, 640, 615, 579, 538, 524, 509, 499, 490, 456, 442, 432 cm^{-1}. – **MS** (FAB, 3-NBA), *m/z*: 472 [M+H]$^+$, 471 [M]$^+$. – **HRMS** (FAB, 3-NBA, C$_{32}$H$_{25}$O$_3$N$_1$) calc.: 471.1834; found: 471.1834.

10-(2'',6-Dimethoxy-[1,1':3',1''-terphenyl]-3-yl)-9,9-dimethyl-9,10-dihydroacridine (60)

A 100 mL round flask was charged with 10-(3-chloro-4-methoxyphenyl)-9,9-dimethyl-9,10-dihydroacridine (1.46 g, 4.17 mmol, 1.00 equiv.), 2-(2'-methoxy-[1,1'-biphenyl]-3-yl)-4,4,5,5-tetramethyl-1,3,2-dioxaborolane (1.43 g, 4.61 mmol, 1.10 equiv.), palladium(II) acetate (46.4 mg, 207 μmol, 5 mol%), 2-dicyclohexylphosphino-2',6'-dimethoxybiphenyl (70.9 mg, 173 μmol, 4 mol%) and potassium phosphate tribasic (2.65 g, 12.5 mmol, 2.99 equiv.). It was evacuated and flushed with argon three times. Through the septum 50 mL toluene was added, then it was heated to 100 °C and stirred for 12 h. The reaction mixture was diluted with 50 mL of ethyl acetate and washed with brine (2 × 50 mL). The organic layer was dried over MgSO$_4$ and the solvent was removed under reduced pressure. The obtained crude product was purified *via* flash-chromatography on silica gel (cyclohexane/dichloromethane = 5:1 to 2:1) to yield the product as a white solid (1.91 g, 3.84 mmol, 92%).

$\mathbf{R}_f$ = 0.35 (cyclohexane/dichloromethane = 5:1). – **^{1}H NMR** (400 MHz, CDCl$_3$) δ = 7.71 (t, J = 1.5 Hz, 1H), 7.55 (dt, J = 7.5 Hz, J = 1.5 Hz, 1H), 7.51 (dt, J = 7.7 Hz, J = 1.5 Hz, 1H), 7.44 (dd, J = 7.6 Hz, J = 1.6 Hz, 3H), 7.38–7.33 (m, 2H), 7.33–7.27 (m, 1H), 7.26–7.17 (m, 2H), 7.04–6.95 (m, 4H), 6.91 (td, J = 7.4 Hz, J = 1.4 Hz, 2H), 6.42 (dd, J = 8.2, J = 1.1 Hz, 2H), 3.93 (s, 3H), 3.79 (s, 3H), 1.68 (s, 6H) ppm. – **^{13}C NMR** (101 MHz, CDCl$_3$) δ = 156.6 (C$_q$, C$_{Ar}$), 156.2 (C$_q$, C$_{Ar}$), 141.4 (C$_q$, 2C$_{Ar}$), 138.5 (C$_q$, C$_{Ar}$), 137.4 (C$_q$, C$_{Ar}$), 133.9 (C$_q$, C$_{Ar}$), 133.7 (+, C$_{Ar}$H), 133.3 (C$_q$, C$_{Ar}$), 131.1 (+, C$_{Ar}$H), 131.0 (+, C$_{Ar}$H), 130.8 (C$_q$, 2C$_{Ar}$), 130.7 (+, C$_{Ar}$H), 130.1 (+, C$_{Ar}$H), 128.7 (+, C$_{Ar}$H), 128.2 (+, 2C$_{Ar}$H), 127.8 (+, C$_{Ar}$H), 126.5 (+, 2C$_{Ar}$H), 125.2 (+, 2C$_{Ar}$H), 120.9 (+, C$_{Ar}$H), 120.5 (+, 2C$_{Ar}$H), 114.2 (+, 2C$_{Ar}$H), 113.1 (+, C$_{Ar}$H), 111.3 (+, C$_{Ar}$H), 56.0 (+, CH$_3$O), 55.7 (+, CH$_3$O), 36.1 (C$_q$, CC$_{Ar}$ (CH$_3$)$_2$), 31.3 (+, 2CH$_3$) ppm. – **IR** (ATR, ṽ) = 2924, 1589, 1499, 1472, 1463, 1443, 1400, 1383, 1324, 1264, 1239, 1179, 1163, 1132, 1122, 1088, 1047, 1026, 928, 902, 798, 744, 704, 663, 618, 602, 579, 431 cm^{-1}. – **MS** (FAB, 3-NBA), *m/z*: 499 [M+H]$^+$, 498 [M]$^+$. – **HRMS** (FAB, 3-NBA, C$_{35}$H$_{32}$O$_2$N$_1$) calc.: 498.2433 [M+H]$^+$; found: 498.2431 [M+H]$^+$.

10,10'-(6,6''-Dimethoxy-[1,1':3',1''-terphenyl]-3,3''-diyl)bis(10H-phenoxazine) (61)

A 50 mL vial was charged with 10-(3-chloro-4-methoxyphenyl)-10H-phenoxazine (906 mg, 2.80 mmol, 1.00 equiv.), 1,3-bis(4,4,5,5-tetramethyl-1,3,2-dioxaborolan-2-yl)benzene (440 mg, 1.33 mmol, 0.95 equiv.), palladium(II) acetate (14.9 mg, 66.4 µmol, 5 mol%), 2-dicyclohexylphosphino-2',6'-dimethoxybiphenyl (54.7 mg, 133 µmol, 10 mol%) and potassium phosphate tribasic (1.70 g, 8.01 mmol, 5.72 equiv.). It was evacuated and flushed with argon three times. Through the septum 20 mL toluene was added, then it was heated to 100 °C and stirred for 12 h. The reaction mixture was diluted with 30 mL of ethyl acetate and washed with brine (2 × 30 mL). The organic layer was dried over MgSO$_4$ and the solvent was removed under reduced pressure. The obtained crude product was purified *via* flash-chromatography on silica gel (cyclohexane/dichloromethane = 10:3 to 2:1) to yield the product as a white solid (600 mg, 919 µmol, 66%).

R$_f$ = 0.40 (cyclohexane/dichloromethane = 10:3). – ^{1}H NMR (400 MHz, CDCl$_3$) δ = 7.69 (t, *J* = 1.6 Hz, 1H), 7.56–7.53 (m, 2H), 7.43-7.47 (m, 1H), 7.34 (d, *J* = 2.6 Hz, 2H), 7.28 (dd, *J* = 8.6 Hz, *J* = 2.6 Hz, 2H), 7.16 (d, *J* = 8.6 Hz, 2H), 6.68–6.57 (m, 12H), 6.03 (d, *J* = 7.3 Hz, 4H), 3.89 (s, 6H) ppm. – ^{13}C NMR (101 MHz, CDCl$_3$) δ = 156.4 (C$_q$, 2C$_{Ar}$OCH$_3$), 144.1 (C$_q$, 2C$_{Ar}$), 137.4 (C$_q$, 4C$_{Ar}$), 134.8 (C$_q$, 2C$_{Ar}$), 133.4 (C$_q$, 4C$_{Ar}$), 133.1 (+, C$_{Ar}$H), 131.6 (C$_q$, 2C$_{Ar}$), 130.8 (+, C$_{Ar}$H), 130.5 (+, 2C$_{Ar}$H), 128.6 (+, 4C$_{Ar}$H), 128.0 (+, 2C$_{Ar}$H), 123.4 (+, 4C$_{Ar}$H), 121.3 (+, 2C$_{Ar}$H), 115.5 (+, 2C$_{Ar}$H), 113.4 (+, 4C$_{Ar}$H), 113.4 (+, 4C$_{Ar}$H), 56.0 (+, 2CH$_3$O) ppm. – IR (ATR, ṽ) = 3033, 1589, 1482, 1462, 1442, 1400, 1330, 1292, 1269, 1245, 1237, 1204, 1181, 1154, 1129, 1116, 1098, 1086, 1068, 1041, 1030, 928, 902, 887, 861, 816, 796, 739, 713, 697, 686, 667, 656, 636, 578, 462, 439, 404 cm^{-1}. – MS (FAB, 3-NBA), *m/z*: 653 [M+H]$^+$, 652 [M]$^+$. – HRMS (FAB, 3-NBA, C$_{44}$H$_{32}$O$_4$N$_2$) calc.: 652.2362; found: 652.2360.

10,10'-(6,6''-Dimethoxy-[1,1':3',1''-terphenyl]-3,3''-diyl)bis(9,9-dimethyl-9,10 dihydroacridine) (62)

A 20 mL vial was charged with 10-(3-chloro-4-methoxyphenyl)-9,9-dimethyl-9,10-dihydroacridine (160 mg, 457 μmol, 1.00 equiv.), 1,3-bis(4,4,5,5-tetramethyl-1,3,2-dioxaborolan-2-yl)benzene (73.6 mg, 223 μmol, 0.98 equiv.), palladium(II) acetate (2.5 mg, 11.1 μmol, 5 mol%), 2-dicyclohexylphosphino-2',6'-dimethoxybiphenyl (9.1 mg, 22.2 μmol, 10 mol%) and potassium phosphate tribasic (284 mg, 1.34 mmol, 5.85 equiv.). It was evacuated and flushed with argon three times. Through the septum 5 mL toluene was added, then it was heated to 100 °C and stirred for 12 h. The reaction mixture was diluted with 20 mL of ethyl acetate and washed with brine (2 × 30 mL). The organic layer was dried over MgSO$_4$ and the solvent was removed under reduced pressure. The obtained crude product was purified *via* flash-chromatography on silica gel (cyclohexane/dichloromethane = 3:1 to 2:1) to yield the product as a white solid (155 mg, 220 μmol, 96%).

R_f = 0.30 (cyclohexane/dichloromethane = 2:1). – **^{1}H NMR** (400 MHz, CDCl$_3$) δ = 7.76 (t, J = 1.5 Hz, 1H), 7.54 (dd, J = 7.3 Hz, J = 1.6 Hz, 2H), 7.45 (dd, J = 7.7 Hz, J = 1.5 Hz, 5H), 7.34 (d, J = 2.5 Hz, 2H), 7.30–7.24 (m, 2H), 7.18 (d, J = 8.6 Hz, 2H), 7.05–6.83 (m, 8H), 6.42 (dd, J = 8.1 Hz, J = 1.2 Hz, 4H), 3.91 (s, 6H), 1.68 (s, 12H) ppm. – **^{13}C NMR** (101 MHz, CDCl$_3$) δ = 156.2 (C$_q$, 2C$_{Ar}$), 141.4 (C$_q$, 4C$_{Ar}$), 137.5 (C$_q$, 2C$_{Ar}$), 133.9 (C$_q$, 2C$_{Ar}$), 133.7 (+, 2C$_{Ar}$H), 133.2 (C$_q$, 2C$_{Ar}$), 131.1 (+, 2C$_{Ar}$H), 130.7 (+, C$_{Ar}$H), 130.1 (C$_q$, 4C$_{Ar}$), 128.5 (+, 2C$_{Ar}$H), 127.9 (+, C$_{Ar}$H), 126.5 (+, 4C$_{Ar}$H), 125.2 (+, 4C$_{Ar}$H), 120.5 (+, 4C$_{Ar}$H), 114.2 (+, 4C$_{Ar}$H), 113.1 (+, 2C$_{Ar}$H), 56.0 (+, 2CH$_3$O), 36.1 (Cq, 2CCAr (CH$_3$)$_2$), 31.3 (+, 4CH$_3$) ppm. – **IR** (ATR, ṽ) = 2919, 2847, 1589, 1496, 1479, 1462, 1441, 1401, 1332, 1282, 1265, 1252, 1241, 1232, 1220, 1180, 1128, 1045, 1026, 803, 748, 741, 704, 602, 507 cm^{-1}. – **MS** (FAB, 3-NBA), *m/z*: 706 [M+H]$^+$, 705 [M]$^+$. – **HRMS** (FAB, 3-NBA, C$_{50}$H$_{45}$O$_2$N$_2$) calc.: 705.3481 [M+H]$^+$; found: 705.3479 [M+H]$^+$.

10-(8,9-Dioxa-8a-borabenzo[fg]tetracen-2-yl)-10H-phenoxazine (45, PXZ-OBO)

Tribromoborane (3.03 g, 12.1 mL, 12.10 mmol, 1.00 M in heptane, 2.00 equiv.) was added to a solution of compound 10-(2,2''-dimethoxy-[1,1':3',1''-terphenyl]-5'-yl)-10H-phenoxazine (2.85 g, 6.04 mmol, 1.00 equiv.) in anhydrous dichlorobenzene (50 mL) under argon. Then the mixture was heated to 150 °C and stirred at this temperature for 12 h. After quenching with methanol, the reaction mixture was concentrated under reduced pressure. The obtained crude product was purified *via* flash-chromatography on silica gel (cyclohexane/dichloromethane = 3:1 to 2:1) to yield the product as a yellow solid (1.50 g, 3.32 mmol, 55%).

R_f = 0.53 (cyclohexane/dichloromethane = 3:1). – **^{1}H NMR** (500 MHz, THF-d_8) δ = 8.30 (m, 4H), 7.59–7.40 (m, 4H), 7.26 (ddd, J = 8.2 Hz, J = 6.6 Hz, J =1.8 Hz, 2H), 6.71 (dd, J = 7.9 Hz, J = 1.6 Hz, 2H), 6.65 (td, J = 7.7 Hz, 1.5 Hz, 2H), 6.57 (td, J = 7.7 Hz, 1.6 Hz, 2H), 6.04 (dd, J = 8.0 Hz, J = 1.5 Hz, 2H) ppm. – **^{13}C NMR** (126 MHz, THF-d_8) δ = 153.2 (C_q, 2C_{Ar}), 146.1 (C_q, 2C_{Ar}), 145.1 (C_q, 2C_{Ar}), 144.2 (C_q, 2C_{Ar}), 135.4 (C_q, 2C_{Ar}), 131.2 (+, 2C_{Ar}H), 125.8 (+, 2C_{Ar}H), 124.4 (+, 2C_{Ar}H), 124.2 (+, 2C_{Ar}H), 123.9 (C_q, C_{Ar}), 123.6 (+, 2C_{Ar}H), 122.5 (+, 2C_{Ar}H), 121.3 (+, 2C_{Ar}H), 116.4 (+, 2C_{Ar}H), 114.6 (+, 2C_{Ar}H) ppm, 1C is missing (1C, C-B).[115] – **^{11}B NMR** (160 MHz, THF-d_8) δ = 26.5 ppm. – **IR** (ATR): $\tilde{v}$ = 3061, 1604, 1581, 1555, 1487, 1463, 1418, 1374, 1339, 1315, 1292, 1269, 1238, 1220, 1207, 1166, 1156, 1143, 1118, 1085, 1044, 1034, 938, 915, 874, 857, 781, 754, 735, 711, 681, 662, 625, 612, 602, 534, 497, 455, 448 cm^{-1}. – **MS** (FAB, 3-NBA), *m/z*: 452 [M+H]$^+$, 451 [M]$^+$. – **HRMS** (FAB, 3-NBA, $C_{30}H_{18}N_1O_3{}^{11}B_1$) calc.: 451.1380; found: 451.1381.

10-(8,9-Dioxa-8a-borabenzo[fg]tetracen-12-yl)-10H-phenoxazine (46, 5PXZ-OBO)

Tribromoborane (988 mg, 3.94 mL, 3.94 mmol, 1.00 M in heptane, 2.00 equiv.) was added to a solution of compound 10-(2'',6-dimethoxy-[1,1':3',1''-terphenyl]-3-yl)-10H-phenoxazine (930 mg, 1.97 mmol, 1.00 equiv.) in anhydrous dichlorobenzene (30 mL) under argon. Then the mixture was heated to 150 °C and stirred at this temperature for 12 h. After quenching with methanol, the reaction mixture was concentrated under reduced pressure. The obtained crude product was purified *via* flash-chromatography on silica gel (cyclohexane/dichloromethane = 2:1 to 1:1) to yield the product as a yellow solid (205 mg, 454 μmol, 23%).

R_f = 0.37 (cyclohexane/dichloromethane = 2:1). – **^{1}H NMR** (300 MHz, THF-d_8) δ = 8.36 (d, J = 2.5 Hz, 1H), 8.34–8.23 (m, 3H), 7.93 (t, J = 7.9 Hz, 1H), 7.68 (d, J = 8.6 Hz, 1H), 7.54–7.40 (m, 3H), 7.29 (ddd, J = 8.3 Hz, J = 6.3 Hz, J = 2.2 Hz, 1H), 6.73–6.51 (m, 6H), 6.01 (dd, J = 7.5 Hz, J = 1.9 Hz, 2H) ppm. – **^{13}C NMR** (101 MHz, THF-d_8) δ = 152.8 (C_q, C_{Ar}), 152.5 (C_q, C_{Ar}), 144.8 (C_q, 2C_{Ar}), 140.3 (C_q, C_{Ar}), 139.5 (C_q, C_{Ar}), 135.4 (C_q, 2C_{Ar}), 134.8 (C_q, C_{Ar}), 134.7 (C_q, C_{Ar}H), 132.6 (C_q, C_{Ar}H), 130.3 (C_q, C_{Ar}H), 127.7 (C_q, C_{Ar}H), 127.0 (C_q, C_{Ar}), 124.9 (C_q, C_{Ar}H), 124.1 (C_q, C_{Ar}), 124.0 (+, 2C_{Ar}H), 123.8 (+, C_{Ar}H), 123.7 (+, C_{Ar}H), 121.9 (+, 2C_{Ar}H), 121.2 (+, C_{Ar}H), 121.1 (+, C_{Ar}H), 120.9 (+, C_{Ar}H), 115.9 (+, 2C_{Ar}H), 114.1 (+, 2C_{Ar}H) ppm, 1C is missing (1C, C-B). – **IR** (ATR): ṽ = 2921, 1604, 1585, 1560, 1483, 1453, 1375, 1333, 1324, 1309, 1292, 1265, 1238, 1204, 1156, 1130, 1116, 1102, 1074, 1060, 1041, 1023, 976, 970, 955, 933, 926, 909, 878, 861, 829, 812, 758, 737, 677, 659, 637, 622, 613, 603, 591, 550, 528, 487, 466, 455, 439, 424, 412, 398, 390, 378 cm^{-1}. – **MS** (FAB, 3-NBA), *m/z*: 452 [M+H]$^+$, 451 [M]$^+$. – **HRMS** ($C_{30}H_{18}N_1O_3{}^{11}B_1$) calc.: 451.1380; found: 451.1378.

5,12-Di(10H-phenoxazin-10-yl)-8,9-dioxa-8a-borabenzo[fg]tetracene (47, DPXZ-OBO)

Tribromoborane (487 mg, 1.94 mL, 1.94 mmol, 1.00 M in heptane, 2.00 equiv.) was added to a solution of compound 10,10'-(6,6''-dimethoxy-[1,1':3',1''-terphenyl]-3,3''-diyl)bis-(10H-phenoxazine) (634 mg, 972 μmol, 1.00 equiv.) in anhydrous dichlorobenzene (30 mL) under argon. Then the mixture was heated to 150 °C and stirred at this temperature for 12 h. After quenching with methanol, the reaction mixture was concentrated under reduced pressure. The obtained crude product was purified *via* flash-chromatography on silica gel (cyclohexane/dichloromethane = 2:1 to 1:1) to yield the product as a yellow solid (128 mg, 202 μmol, 21%).

R_f = 0.58 (cyclohexane/dichloromethane = 1.5:1). – **^{1}H NMR** (500 MHz, THF-d_8) δ = 8.38 (d, J = 2.5 Hz, 2H), 8.31 (d, J = 7.9 Hz, 2H), 7.93 (t, J = 7.9 Hz, 1H), 7.71 (d, J = 8.5 Hz, 2H), 7.49 (dd, J = 8.6 Hz, J = 2.4 Hz, 2H), 6.68 (dd, J = 7.8 Hz, J = 1.6 Hz, 4H), 6.62 (td, J = 7.6 Hz, J = 1.6 Hz, 4H), 6.58 (td, J = 7.6 Hz, J = 1.7 Hz, 4H), 6.03 (dd, J = 7.8 Hz, J = 1.5 Hz, 4H) ppm. – **^{13}C NMR** (101 MHz, THF-d_8) δ = 152.4 (C_q, 2C_{Ar}), 144.8 (C_q, 4C_{Ar}), 139.5 (C_q, 2C_{Ar}), 135.4 (C_q, 4C_{Ar}), 135.0 (C_q, 2C_{Ar}), 134.9 (+, C_{Ar}H), 132.7 (+, 2C_{Ar}H), 127.8 (+, 2C_{Ar}H), 126.9 (C_q, 2C_{Ar}), 124.0 (+, 4C_{Ar}H), 123.8 (+, 2C_{Ar}H), 121.9 (+, 4C_{Ar}H), 121.8 (+, 2C_{Ar}H), 115.9 (+, 4C_{Ar}H), 114.1 (+, 4C_{Ar}H) ppm, 1C is missing (1C, C–B). – **IR** (ATR): ṽ = 2922, 1605, 1588, 1562, 1486, 1456, 1377, 1339, 1330, 1316, 1292, 1265, 1203, 1153, 1119, 1103, 1095, 1075, 1043, 935, 914, 878, 871, 861, 829, 813, 778, 730, 674, 640, 630, 613, 601, 562, 492, 453, 426, 394 cm^{-1}. – **MS** (FAB, 3-NBA), *m/z*: 633 [M+H]$^+$, 632 [M]$^+$. – **HRMS** ($C_{42}H_{25}N_2O_4{}^{11}B_1$) calc.: 632.1907; found: 632.1907.

10-(8,9-Dioxa-8a-borabenzo[fg]tetracen-2-yl)-9,9-dimethyl-9,10-dihydroacridine (48, DMAC-OBO)

Tribromoborane (1.00 g, 4.00 mL, 4.00 mmol, 1.00 M in heptane, 1.99 equiv.) was added to a solution of compound 10-(2,2''-dimethoxy-[1,1':3',1''-terphenyl]-5'-yl)-9,9-dimethyl-9,10-dihydroacridine (1.00 g, 2.01 mmol, 1.00 equiv.) in anhydrous dichlorobenzene (30 mL) under argon. Then the mixture was heated to 150 °C and stirred at this temperature for 12 h. After quenching with methanol, the reaction mixture was concentrated under reduced pressure. The obtained crude product was purified *via* flash-chromatography on silica gel (cyclohexane/dichloromethane = 3:1) to yield the product as a white solid (535 mg, 1.12 mmol, 56%).

R_f = 0.56 (cyclohexane/dichloromethane = 3:1). – **^{1}H NMR** (500 MHz, THF-d_8) δ = 8.28 (dd, J = 8.1 Hz, J =1.2 Hz, 2H), 8.24 (s, 2H), 7.54–7.49 (m, 2H), 7.49–7.43 (m, 4H), 7.23 (ddd, J = 8.2 Hz, J =5.6 Hz, J =2.7 Hz, 2H), 6.94–6.85 (m, 4H), 6.36–6.29 (m, 2H), 1.75 (s, 6H) ppm. – **^{13}C NMR** (126 MHz, THF-d_8) δ = 152.9 (C$_q$, 2C$_{Ar}$), 148.0 (C$_q$, C$_{Ar}$), 143.6 (C$_q$, 2C$_{Ar}$), 141.5 (C$_q$, 2C$_{Ar}$), 130.7 (+, 2C$_{Ar}$H), 130.6 (C$_q$, 2C$_{Ar}$), 127.0 (+, 2C$_{Ar}$H), 126.0 (+, 2C$_{Ar}$H), 125.5 (+, 2C$_{Ar}$H), 123.8 (+, 2C$_{Ar}$H), 123.7 (+, 2C$_{Ar}$H), 123.6 (C$_q$, 2C$_{Ar}$), 121.3 (+, 2C$_{Ar}$H), 120.9 (+, 2C$_{Ar}$H), 114.9 (+, 2C$_{Ar}$H), 36.6 (C$_q$, CC$_{Ar}$(CH$_3$)$_2$), 31.9 (+, 2CH$_3$) ppm, 1C is missing (1C, C–B). –**^{11}B NMR** (160 MHz, THF-d_8) δ = 27.81 ppm. – **IR** (ATR): $\tilde{v}$ = 2968, 1611, 1589, 1584, 1568, 1555, 1500, 1480, 1460, 1438, 1418, 1371, 1334, 1281, 1258, 1218, 1205, 1174, 1157, 1146, 1135, 1119, 1085, 1044, 1035, 972, 941, 933, 916, 874, 856, 807, 779, 764, 744, 737, 704, 674, 632, 619, 611, 572, 533, 497, 453, 428 cm^{-1}. – **MS** (EI), *m/z*: 478 [M+H]$^+$, 477 [M]$^+$. – **HRMS** (EI, C$_{33}$H$_{24}$N$_1$O$_2$^{11}B$_1$) calc.: 477.1900; found: 477.1898.

10-(8,9-Dioxa-8a-borabenzo[fg]tetracen-12-yl)-9,9-dimethyl-9,10-dihydroacridine   **(49,**
5DMAC-OBO)

Tribromoborane (1.71 g, 6.83 mL, 6.83 mmol, 1.00 M in heptane, 2.00 equiv.) was added to a solution of compound 10-(2'',6-dimethoxy-[1,1':3',1''-terphenyl]-3-yl)-9,9-dimethyl-9,10-dihydroacridine (1.70 g, 3.42 mmol, 1.00 equiv.) in anhydrous dichlorobenzene (45 mL) under argon. Then the mixture was heated to 150 °C and stirred at this temperature for 12 h. After quenching with methanol, the reaction mixture was concentrated under reduced pressure. The obtained crude product was purified *via* flash-chromatography on silica gel (cyclohexane/dichloromethane = 3:1 to 2:1) to yield the product as a white solid (360 mg, 754 μmol, 22%).

R_f = 0.54 (cyclohexane/dichloromethane = 3:1). – **^{1}H NMR** (500 MHz, THF-d_8) δ = 8.32–8.27 (m, 2H), 8.24 (dd, J = 14.4 Hz, J = 8.0 Hz, 2H), 7.90 (t, J = 7.9 Hz, 1H), 7.70 (d, J = 8.5 Hz, 1H), 7.51–7.38 (m, 5H), 7.29 (ddd, J = 8.2 Hz, J = 6.3 Hz, J = 2.0 Hz, 1H), 6.89 (m, 4H), 6.35 (dd, J = 8.1 Hz, J = 1.4 Hz, 2H), 1.70 (s, 6H) ppm. – **^{13}C NMR** (126 MHz, THF-d_8) δ = 153.2 (C_q, C_{Ar}), 152.7 (C_q, C_{Ar}), 142.3 (C_q, C_{Ar}), 140.6 (C_q, C_{Ar}), 140.0 (C_q, C_{Ar}), 137.5 (C_q, C_{Ar}), 135.1 (+, C_{Ar}H), 133.4 (+, C_{Ar}H), 131.0 (C_q, C_{Ar}), 130.6 (+, C_{Ar}H), 128.4 (+, C_{Ar}H), 127.3 (+, C_{Ar}H), 127.1 (C_q, C_{Ar}), 126.1 (+, C_{Ar}H), 125.3 (+, C_{Ar}H), 124.4 (C_q, C_{Ar}), 124.1 (+, C_{Ar}H), 123.8 (+, C_{Ar}H), 121.5 (+, C_{Ar}H), 121.4 (+, C_{Ar}H), 121.3 (+, C_{Ar}H), 115.1 (+, C_{Ar}H), 36.9 (C_q, CCAr (CH$_3$)$_2$), 31.9 (+, CH$_3$) ppm, 1C is missing (1C, C–B). – **IR** (ATR): ṽ = 2955, 2948, 1602, 1587, 1560, 1486, 1473, 1446, 1375, 1329, 1288, 1269, 1241, 1210, 1177, 1162, 1130, 1118, 1103, 1089, 1074, 1061, 1044, 1024, 929, 877, 861, 837, 827, 810, 754, 739, 701, 666, 630, 616, 567, 543, 531, 509, 483, 425 cm^{-1}. – **MS** (FAB, 3-NBA), *m/z*: 478 [M+H]$^+$, 477 [M]$^+$. – **HRMS** (C$_{33}$H$_{24}$N$_1$O$_2$^{11}B$_1$) calc.: 477.1900; found: 477.1901.

5,12-Bis(9,9-dimethylacridin-10(9H)-yl)-8,9-dioxa-8a-borabenzo[fg]tetracene (50, DDMAC-OBO)

Tribromoborane (995 mg, 3.97 mL, 3.97 mmol, 1.00 M in heptane, 2.00 equiv.) was added to a solution of compound 10,10'-(6,6''-dimethoxy-[1,1':3',1''-terphenyl]-3,3''-diyl)bis-(9,9-dimethyl-9,10-dihydr-oacridine) (1.40 g, 1.99 mmol, 1.00 equiv.) in anhydrous dichlorobenzene (50 mL) under argon. Then the mixture was heated to 150 °C and stirred at this temperature for 12 h. After quenching with methanol, the reaction mixture was concentrated under reduced pressure. The obtained crude product was purified *via* flash-chromatography on silica gel (cyclohexane/dichloromethane = 2:1 to 1:1) to yield the product as a white solid (320 mg, 467 μmol, 24%).

R$_f$ = 0.28 (cyclohexane/dichloromethane = 2:1). – **^{1}H NMR** (500 MHz, THF-d_8) δ = 8.32 (d, J = 2.4 Hz, 2H), 8.27 (d, J = 8.0 Hz, 2H), 7.88 (t, J = 7.9 Hz, 1H), 7.75 (d, J = 8.5 Hz, 2H), 7.48 (dd, J = 7.7 Hz, J = 1.6 Hz, 4H), 7.46 (dd, J = 8.5 Hz, J = 2.4 Hz, 2H), 6.90 (m, 8H), 6.37 (dd, J = 8.2 Hz, J = 1.3 Hz, 4H), 1.70 (s, 12H) ppm. – **^{13}C NMR** (126 MHz, THF-d_8) δ = 152.3 (C$_q$, 2C$_{Ar}$), 141.9 (C$_q$, 4C$_{Ar}$), 139.6 (C$_q$, 2C$_{Ar}$), 137.3 (C$_q$, 2C$_{Ar}$), 134.8 (+, C$_{Ar}$H), 133.2 (+, 2C$_{Ar}$H), 130.7 (C$_q$, 4C$_{Ar}$), 128.1 (+, 2C$_{Ar}$H), 126.9 (+, 4C$_{Ar}$H), 126.7 (+, 2C$_{Ar}$H), 125.8 (+, 4C$_{Ar}$H), 123.5 (+, 2C$_{Ar}$H), 121.7 (+, 2C$_{Ar}$H), 121.2 (+, 4C$_{Ar}$H), 114.7 (+, 4C$_{Ar}$H), 36.6 (C$_q$, 2CCAr (CH$_3$)$_2$), 31.5 (+, 4CH$_3$) ppm, 1C is missing (1C, C–B). – **IR** (ATR): ṽ = 2925, 1591, 1558, 1499, 1473, 1460, 1449, 1378, 1329, 1316, 1264, 1242, 1211, 1173, 1120, 1072, 1047, 929, 834, 815, 739, 652, 629, 596, 579, 517, 487, 469, 428 cm^{-1}. – **MS** (FAB, 3-NBA), *m/z*: 685 [M+H]$^+$, 684 [M]$^+$. – **HRMS** (C$_{48}$H$_{37}$N$_2$O$_2$^{11}B$_1$) calc.: 684.2948; found: 684.2948.

3,6-Di-*tert*-butyl-9-(2,5-dibromophenyl)-9*N*-carbazole (72)

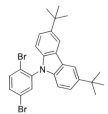

Under argon atmosphere, dry DMF (50 mL) was injected to the mixture of 3,6-ditert-butyl-9*H*-carbazole (2.46 g, 8.80 mmol, 1.10 equiv.) and sodium hydride (382 mg, 15.9 mmol, 1.99 equiv.), then the reaction mixture was stirred at room temperature for 1 h. 1,4-dibromo-2-fluorobenzene (2.03 g, 8.00 mmol, 1.00 equiv.) was added to the reaction mixture and stirred at 100 °C for 12 h. After cooling to room temperature, the reaction mixture was diluted with 100 mL of DCM and washed with brine (3 × 50 mL). The organic layer was dried over $MgSO_4$ and the solvent was removed under reduced pressure. The obtained crude product was purified *via* column chromatography on silica gel (cyclohexane/ethyl acetate = 50:1) to yield the title compound as a white solid (3.79 g, 7.38 mmol, 92%).

R_f = 0.60 (cyclohexane/ethyl acetate = 50:1). – **^{1}H NMR** (400 MHz, CDCl$_3$) δ = 8.16 (d, J = 1.7 Hz, 2H), 7.72 (d, J = 8.6 Hz, 1H), 7.60 (d, J = 2.4 Hz, 1H), 7.52 (dd, J = 8.6 Hz, J = 2.4 Hz, 1H), 7.48 (dd, J = 8.6 Hz, J = 1.9 Hz, 2H), 7.02 (dd, J = 8.6 Hz, J = 0.6 Hz, 2H), 1.49 (s, 18H) ppm. – **^{13}C NMR** (101 MHz, CDCl$_3$) δ = 143.4 (C_q, C_{Ar}), 139.2 (C_q, C_{Ar}), 138.9 (C_q, C_{Ar}), 135.3 (+, C_{Ar}H), 134.1 (+, C_{Ar}H), 133.0 (+, C_{Ar}H), 123.9 (+, C_{Ar}H), 123.5 (C_q, C_{Ar}), 122.7 (C_q, C_{Ar}), 121.7 (C_q, C_{Ar}), 116.6 (+, C_{Ar}H), 109.6 (+, C_{Ar}H), 34.9 (C_q, C(CH$_3$)$_3$), 32.2 (+, CH_3) ppm. – **IR** (ATR, ṽ) = 2949 (s), 2927 (w), 2900 (w), 2861 (w), 1571 (w), 1482 (vs), 1459 (s), 1452 (m), 1404 (w), 1392 (w), 1363 (m), 1323 (w), 1293 (s), 1258 (s), 1232 (s), 1137 (w), 1075 (m), 1035 (m), 962 (w), 878 (m), 816 (s), 805 (vs), 739 (w), 649 (w), 615 (s), 545 (w), 528 (w), 452 (m), 426 (w) cm^{-1}. – **MS** (FAB, 3-NBA), *m/z* (%): 513 [M]$^+$. – **HRMS** (FAB, 3-NBA) calc. for C$_{26}$H$_{27}$N$_1$^{79}Br$_1$^{81}Br$_1$ [M]$^+$ 513.0490, found 513.0488.

(s)-9-(5-Bromo-2-(dimesitylboraneyl)phenyl)-3,6-di-tert-butyl-9*N*-carbazole (74)

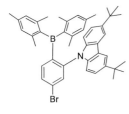

n-BuLi in hexane (440 μL, 1.10 mmol, 2.50M, 1.10 equiv.) was added dropwise to a stirred solution of 3,6-di-tert-butyl-9-(2,5-dibromophenyl)-9*N*-carbazole (513 mg, 1000 μmol, 1.00 equiv.) in dry Et$_2$O (25 mL) at –78 °C under argon atmosphere. The mixture was stirred at –78 °C for 1.5 h. Dimesitylboron fluoride (322 mg, 1.20 mmol, 1.20 equiv.) in Et$_2$O (5 mL) was added to this mixture at 0 °C. After stirring for 12 h at room temperature, the reaction mixture was poured into saturated aqueous NH$_4$Cl solution and extracted with 25 mL DCM. Then it was washed with brine (3 × 25 mL). The organic layer was dried over MgSO$_4$ and the solvent was removed under reduced pressure. The obtained crude product was purified *via* column chromatography on silica gel (cyclohexane = 1) to yield the title compound as a white solid (170 mg, 249 μmol, 25%).

R$_f$ = 0.70 (cyclohexane = 1). – **^{1}H NMR** (400 MHz, CDCl$_3$) δ = 7.78 (d, *J* = 1.9 Hz, 2H), 7.71–7.61 (m, 2H), 7.55 (d, *J* = 8.1 Hz, 1H), 7.33 (dd, *J* = 8.6, 1.9 Hz, 2H), 7.15–7.06 (m, 2H), 6.81–5.60 (br, 4H, MesH), 2.51–0.70 (m, 18H), 1.44 (s, 18H) ppm.[126-127] – **IR** (ATR, ṽ) = 2953 (s), 2915 (m), 2907 (m), 2864 (w), 1606 (m), 1567 (s), 1538 (w), 1486 (vs), 1475 (vs), 1453 (s), 1418 (s), 1397 (s), 1361 (s), 1326 (w), 1295 (s), 1259 (vs), 1235 (vs), 1218 (m), 1201 (s), 1169 (w), 1154 (m), 1135 (w), 1033 (w), 962 (m), 894 (w), 875 (m), 861 (w), 840 (vs), 810 (vs), 721 (m), 691 (w), 662 (m), 613 (s) cm^{-1}. – **MS** (FAB, 3-NBA), *m/z* (%): 681 [M]$^+$. – **HRMS** (FAB, 3-NBA) calc. for C$_{44}$H$_{49}$N$_1$^{11}B$_1$^{79}Br$_1$ [M]$^+$ 681.3141, found 681.3140.

9-(5-(9N-Carbazol-9-yl)-2-(dimesitylboraneyl)phenyl)-3,6-di-tert-butyl-9N-carbazole (68)

Under argon atmosphere, a mixture of carbazole (16.7 mg, 99.9 μmol, 1.00 equiv.), (s)-9-(5-bromo-2-(dimesitylboraneyl)phenyl)-3,6-di-tert-butyl-9N-carbazole (71.6 mg, 105 μmol, 1.05 equiv.), Pd$_2$(dba)$_3$ (4.6 mg, 5.02 μmol, 5 mol%), tri-*tert*-butylphosphonium tetrafluoroborate (2.9 mg, 10.00 μmol, 10 mol%) and sodium *tert*-butoxide (48.1 mg, 501 μmol, 5.01 equiv.) in toluene (4.0 mL) was stirred at 100 °C for 12 h. The mixture was cooled to room temperature and diluted with dichloromethane (20 mL). Then it was washed with brine (3 × 20 mL). The organic layer was dried over MgSO$_4$ and the solvent was removed under reduced pressure. The obtained crude product was purified *via* column chromatography on silica gel (cyclohexane/dichloromethane = 5:1) to yield the title compound as a yellow solid (65.0 mg, 84.5 μmol, 85%).

R_f = 0.40 (cyclohexane/ethyl acetate = 50:1). – **^{1}H NMR** (400 MHz, CDCl$_3$) δ = 8.15 (d, J = 7.7 Hz, 2H), 7.97 (d, J = 8.1 Hz, 1H), 7.83–7.73 (m, 4H), 7.65 (d, J = 8.2 Hz, 2H), 7.51–7.42 (m, 2H), 7.38–7.28 (m, 4H), 7.25 (d, J = 8.2 Hz, 2H), 7.14–5.42 (br, 4H, MesH), 2.98–0.70 (m, 18H), 1.45 (s, 18H) ppm. – **^{13}C NMR** (101 MHz, CDCl$_3$) δ = 145.4 (C$_q$, C$_{Ar}$), 142.3 (C$_q$, C$_{Ar}$), 140.6 (+, C$_{Ar}$H), 140.2 (C$_q$, C$_{Ar}$), 127.5 (+, C$_{Ar}$H), 126.4 (+, C$_{Ar}$H), 125.4 (+, C$_{Ar}$H), 124.1 (C$_q$, C$_{Ar}$), 123.4 (C$_q$, C$_{Ar}$), 122.6 (+, C$_{Ar}$H), 120.8 (+, C$_{Ar}$H), 120.6 (+, C$_{Ar}$H), 115.4 (+, C$_{Ar}$H), 110.1 (+, C$_{Ar}$H), 109.7 (+, C$_{Ar}$H), 34.8 (C$_q$, C(CH$_3$)$_3$), 32.3 (+, CH$_3$), 32.1 (+, CH$_3$), 21.3 (+, CH$_3$) ppm. – **IR** (ATR, ṽ) = 2953 (m), 1605 (m), 1589 (s), 1476 (s), 1446 (vs), 1418 (s), 1361 (s), 1333 (m), 1309 (w), 1293 (m), 1261 (s), 1227 (s), 1203 (s), 1164 (m), 1154 (m), 840 (s), 809 (s), 747 (vs), 721 (vs), 710 (m), 662 (m), 615 (m), 422 (m) cm^{-1}. – **MS** (FAB, 3-NBA), m/z (%): 769 [M+H]$^+$, 768 [M]$^+$. – **HRMS** (FAB, 3-NBA) calc. for C$_{56}$H$_{57}$N$_2$^{11}B$_1$ [M]$^+$ 768.4615, found 768.4614.

5,11-Bis(3-(3,6-di-tert-butyl-9N-carbazol-9-yl)-4-(dimesitylboraneyl)phenyl)-5,11-dihydroindolo[3,2-b]carbazole (69)

Under argon atmosphere, a mixture of 5,11-dihydroindolo[3,2-b]carbazole (12.8 mg, 49.9 μmol, 1.00 equiv.), (s)-9-(5-bromo-2-(dimesitylboraneyl)phenyl)-3,6-di-tert-butyl-9N-carbazole (71.6 mg, 105 μmol, 2.10 equiv.), Pd$_2$(dba)$_3$ (4.6 mg, 5.02 μmol, 10 mol%), tri-tert-butylphosphonium tetrafluoroborate (2.9 mg, 10.00 μmol, 20 mol%) and sodium tert-butoxide (48.1 mg, 501 μmol, 10.00 equiv.) in toluene (4.0 mL) was stirred at 100 °C for 12 h. Then the mixture was cooled to room temperature and diluted with dichloromethane (20 mL). Then it was washed with brine (3 × 20 mL). The organic layer was dried over MgSO$_4$ and the solvent was removed under reduced pressure. The obtained crude product was purified *via* column chromatography on silica gel (cyclohexane/dichloromethane = 15:1 to 5:1) to yield the title compound as a yellow solid (45.0 mg, 30.8 μmol, 62%).

R$_f$ = 0.40 (cyclohexane/ethyl acetate = 50:1). – **^{1}H NMR** (400 MHz, CDCl$_3$) δ = 8.39 (bs, 4H), 8.20 (d, J = 7.7 Hz, 2H), 8.01 (d, J = 8.1 Hz, 2H), 7.93 (d, J = 2.0 Hz, 4H), 7.88 (dd, J = 8.1 Hz, J = 2.1 Hz, 2H), 7.80 (bs, 4H), 7.72 (d, J = 8.3 Hz, 2H), 7.50–7.39 (m, 4H), 7.38–7.27 (m, 4H), 7.08–5.56 (br, 8H), 2.82–0.70 (m, 36H), 1.47 (s, 36H) ppm. – **^{13}C NMR** (101 MHz, CDCl$_3$) δ = 145.4 (C$_q$, C$_{Ar}$), 143.5 (C$_q$, C$_{Ar}$), 142.8 (C$_q$, C$_{Ar}$), 142.3 (C$_q$, C$_{Ar}$), 141.2 (C$_q$, C$_{Ar}$), 140.6 (+, C$_{Ar}$H), 136.4 (C$_q$, C$_{Ar}$), 127.3 (+, C$_{Ar}$H), 126.6 (+, C$_{Ar}$H), 125.5 (C$_q$, C$_{Ar}$), 125.0 (+, C$_{Ar}$H), 124.3 (C$_q$, C$_{Ar}$), 124.1 (C$_q$, C$_{Ar}$), 123.5 (C$_q$, C$_{Ar}$), 122.7 (+, C$_{Ar}$H), 120.7 (+, C$_{Ar}$H), 120.5 (+, C$_{Ar}$H), 115.7 (C$_q$, C$_{Ar}$), 115.4 (+, C$_{Ar}$H), 110.1 (+, C$_{Ar}$H), 109.7 (+, C$_{Ar}$H), 100.6 (+, C$_{Ar}$H), 97.5 (C$_q$, C$_{Ar}$), 34.8 (C$_q$, *C*(CH$_3$)$_3$), 32.3 (+, *C*H$_3$), 21.3 (+, *C*H$_3$) ppm. – **IR** (ATR, ṽ) = 2961 (m), 1606 (m), 1587 (vs), 1476 (vs), 1443 (vs), 1422 (s), 1363 (m), 1319 (s), 1295 (m), 1255 (vs), 1234 (vs), 1215 (s), 1203 (vs), 1191 (s), 1170 (m), 1162 (s), 841 (vs), 830 (s), 807 (s), 759 (m), 741 (vs), 721 (m), 663 (m), 612 (w) cm^{-1}. – **MS** (FAB, 3-NBA), *m/z* (%): 1460 [M+H]$^+$, 1459 [M]$^+$.

5.3 Analytical Data of Indolocarbazole-Based Donors for TADF Emitters

2,4-Bis(4-(*tert*-butyl)phenyl)-6-chloro-1,3,5-triazine (91)

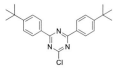

A solution of 1-bromo-4-(*tert*-butyl)benzene (4.00 g, 18.8 mmol, 2.16 equiv.) in anhydrous THF (10 ml) was added dropwise to a suspension of iodine-activated magnesium (2.30 g, 94.7 mmol, 10.80 equiv.) in anhydrous THF (5 mL). After that the reaction mixture was refluxed for 2 h and then cooled to room temperature. The resulting Grignard solution was slowly added to a solution of cyanuric chloride (1.60 g, 8.68 mmol, 1.00 equiv.) in anhydrous toluene (50 mL) at room temperature. The mixture was refluxed for 3 h and was cooled down to room temperature. The reaction mixture was quenched with distilled water and extracted with ethyl acetate. The product was obtained through recrystallization from MeOH as a white solid (1.93 g, 5.07 mmol, 58%).

R_f = 0.45 (cyclohexane/ethyl acetate = 50:1). – ^{1}H NMR (400 MHz, CDCl$_3$) δ = 8.58–8.50 (m, 4H), 7.59–7.53 (m, 4H), 1.39 (s, 18H) ppm. – ^{13}C NMR (101 MHz, CDCl$_3$) δ = 173.4 (C$_q$, 2C$_{Ar}$), 172.1 (C$_q$, C$_{Ar}$), 157.5 (C$_q$, 2C$_{Ar}$), 131.9 (C$_q$, 2C$_{Ar}$), 129.4 (+, 4C$_{Ar}$H), 125.9 (+, 4C$_{Ar}$H), 35.4 (C$_q$, 2C(CH$_3$)$_3$), 31.3 (+, 6CH$_3$) ppm. – IR (ATR, ṽ) = 2963 (w), 2952 (w), 1608 (w), 1571 (w), 1524 (vs), 1482 (vs), 1412 (w), 1394 (w), 1368 (vs), 1322 (m), 1313 (m), 1299 (m), 1265 (w), 1245 (vs), 1193 (m), 1162 (w), 1105 (w), 1075 (m), 1018 (w), 850 (s), 829 (m), 807 (vs), 722 (m), 703 (m), 659 (w), 545 (s) cm^{-1}. – MS (EI, 70 eV), *m/z*: 380 [M+H]$^+$, 379 [M]$^+$. – HRMS (FAB, 3-NBA) calc. for C$_{23}$H$_{27}$N$_3$^{35}Cl$_1$ [M+H]$^+$ 380.1894, found 380.1891.

2-Chloro-4,6-bis(3,5-di-*tert*-butylphenyl)-1,3,5-triazine (92)

Under argon atmosphere, a solution of 1-bromo-3,5-di-*tert*-butylbenzene (8.07 g, 30.0 mmol, 2.17 equiv.) in anhydrous THF (30 mL) was added dropwise to a suspension of iodine-activated magnesium (3.66 g, 150.6 mmol, 10.91 equiv.) in anhydrous THF (15 mL). After that the reaction mixture was refluxed for 2 h and then cooled to room temperature. The resulting Grignard solution was slowly added to a solution of cyanuric chloride (2.55 g, 13.8 mmol, 1.00 equiv.) in anhydrous toluene (45 mL) at room temperature. The mixture was refluxed for 3 h and was cooled down to room temperature. The reaction mixture was quenched with distilled water and extracted with ethyl acetate. The product was obtained through recrystallization from MeOH as a white solid (4.26 g, 8.66 mmol, 63%).

R_f = 0.60 (cyclohexane/ethyl acetate = 50:1). – ^{1}H NMR (400 MHz, CDCl$_3$) δ = 8.52 (d, J = 1.9 Hz, 4H), 7.72 (t, J = 1.9 Hz, 2H), 1.44 (s, 18H) ppm. – ^{13}C NMR (101 MHz, CDCl$_3$) δ = 173.9 (C$_q$, C$_{Ar}$), 172.3 (C$_q$, C$_{Ar}$), 151.6 (C$_q$, C$_{Ar}$), 134.0 (C$_q$, C$_{Ar}$), 128.0 (+, C$_{Ar}$H), 123.7 (+, C$_{Ar}$H), 35.2 (C$_q$, C(CH$_3$)$_3$), 31.5 (+, CH$_3$) ppm. – IR (ATR, ṽ) = 2959 (m), 2902 (w), 2866 (w), 1523 (vs), 1499 (vs), 1477 (m), 1460 (s), 1425 (m), 1392 (w), 1383 (w), 1364 (vs), 1317 (w), 1286 (vs), 1244 (vs), 1203 (w), 1169 (w), 1139 (w), 1125 (w), 1106 (w), 922 (w), 892 (s), 824 (vs), 734 (s), 705 (s), 686 (s) cm^{-1}. – MS (EI, 70 eV), m/z (%): 492 [M+H]$^+$, 491 [M]$^+$. – HRMS (FAB, 3-NBA) calc. for C$_{31}$H$_{43}$O$_2$N$_3$^{35}Cl$_1$ [M+H]$^+$ 492.3146, found 492.3144.

2,4-Bis(4-(tert-butyl)phenyl)-6-(4-fluorophenyl)-1,3,5-triazine (93)

Under argon atmosphere, a mixture of 2,4-bis(4-tert-butylphenyl)-6-chloro-1,3,5-triazine (2.52 g, 6.63 mmol, 1.00 equiv.), (4-fluorophenyl)boronic acid (0.99 g, 7.08 mmol, 1.07 equiv.), tetrakis(triphenylphosphine)palladium(0) (0.39 g, 337 μmol, 5 mol%) and aqueous potassium carbonate (4M, 8.5 mL, 34.00 mmol, 5.12 equiv.) in PhMe/EtOH (60.0/8.5 mL) was stirred at 90 °C for 2.5 h. After cooling to room temperature, the mixture was diluted with dichloromethane (50 mL). Then it was washed with brine (2 × 50 mL), dried over MgSO$_4$ and the solvent was removed under reduced pressure. The obtained crude product was purified *via* column chromatography on silica gel (cyclohexane/ethyl acetate = 50:1 to 30:1) to yield the title compound as a white solid (2.55 g, 5.80 mmol, 87%).

R_f = 0.40 (cyclohexane/ethyl acetate = 50:1). – ^{1}H NMR (400 MHz, CDCl$_3$) δ = 8.83–8.73 (m, 2H), 8.71–8.63 (m, 4H), 7.65–7.55 (m, 4H), 7.29–7.17 (m, 2H), 1.42 (s, 18H) ppm. – ^{13}C NMR (100 MHz, CDCl$_3$) δ = 171.7 (C$_q$, 2C$_{Ar}$), 170.5 (C$_q$, C$_{Ar}$), 165.8 (C$_q$, d, $^1J_{CF}$ = 252.6 Hz, C$_{Ar}$), 156.2 (C$_q$, 2C$_{Ar}$), 133.7 (C$_q$, 2C$_{Ar}$), 132.8 (d, $^4J_{CF}$ = 2.8 Hz), 131.3 (+, d, $^3J_{CF}$ = 9.0 Hz, 2C$_{Ar}$H), 128.9 (+, 4C$_{Ar}$H), 125.7 (+, 4C$_{Ar}$H), 115.7 (+, d, $^2J_{CF}$ = 21.6 Hz, 2C$_{Ar}$H), 35.2 (C$_q$, 2C(CH$_3$)$_3$), 31.4 (+, 6CH$_3$) ppm. – ^{19}F NMR (376 MHz, CDCl$_3$) δ = –107.5 ppm. – IR (ATR, ṽ) = 2961 (w), 1601 (s), 1578 (w), 1517 (vs), 1503 (vs), 1460 (m), 1411 (s), 1367 (vs), 1356 (vs), 1305 (w), 1289 (w), 1265 (m), 1222 (s), 1204 (w), 1191 (m), 1145 (s), 1108 (w), 1095 (w), 1014 (m), 858 (m), 813 (vs), 704 (w), 694 (w), 581 (s), 567 (w), 547 (s), 504 (m) cm^{-1}. – MS (EI, 70 eV), *m/z*: 440 [M+H]$^+$, 439 [M]$^+$. – HRMS (FAB, 3-NBA) calc. for C$_{29}$H$_{31}$N$_3$F$_1$ [M+H]$^+$ 440.2502, found 440.2504.

2,4-Bis(3,5-di-*tert*-butylphenyl)-6-(4-fluorophenyl)-1,3,5-triazine (94)

Under argon atmosphere, a mixture of 2-chloro-4,6-bis(3,5-di-*tert*-butylphenyl)-1,3,5-triazine (328 mg, 667 μmol, 1.00 equiv.), (4-fluorophenyl)boronic acid (97.9 mg, 700 μmol, 1.05 equiv.), tetrakis(triphenylphosphine)palladium(0) (38.5 mg, 33.3 μmol, 5 mol%) and aqueous potassium carbonate (4M, 0.8 mL, 3.20 mmol, 4.80 equiv.) in PhMe/EtOH (6.7/0.8 mL) was stirred at 90 °C for 2.5 h. After cooling to room temperature, the reaction mixture was diluted with 20 mL of ethyl acetate and washed with brine (3 × 20 mL). The organic layer was dried over MgSO₄ and the solvent was removed under reduced pressure. The obtained crude product was purified *via* column chromatography on silica gel (cyclohexane/ethyl acetate = 50:1 to 30:1) to yield the title compound as a white solid (350 mg, 634 μmol, 95%).

R_f = 0.30 (cyclohexane/ethyl acetate = 50:1). – **¹H NMR** (400 MHz, CDCl₃) δ = 8.85–8.73 (m, 2H), 8.63 (d, J = 1.9 Hz, 4H), 7.69 (t, J = 1.9 Hz, 2H), 7.30–7.21 (m, 2H), 1.45 (s, 36H) ppm. – **¹³C NMR** (101 MHz, CDCl₃) δ = 172.3 (C_q, C_Ar), 170.6 (C_q, C_Ar), 165.8 (C_q, d, $^1J_{CF}$ = 252.4 Hz, C_Ar), 151.3 (C_q, C_Ar), 135.6 (C_q, C_Ar), 132.9 (C_q, d, $^4J_{CF}$ = 2.8 Hz, C_Ar), 131.4 (+, d, $^3J_{CF}$ = 9.0 Hz, C_ArH), 127.1 (+, C_ArH), 123.3 (+, C_ArH), 115.8 (+, d, $^2J_{CF}$ = 21.8 Hz, C_ArH), 35.2 (C_q, C(CH₃)₃), 31.6 (+, CH₃) ppm. – **¹⁹F NMR** (376 MHz, CDCl₃) δ = –111.81 ppm. – **IR** (ATR, ṽ) = 2951 (m), 2902 (w), 2866 (w), 1599 (w), 1520 (vs), 1509 (vs), 1477 (m), 1460 (m), 1436 (m), 1407 (w), 1391 (w), 1354 (vs), 1300 (w), 1281 (w), 1266 (w), 1247 (m), 1237 (w), 1221 (s), 1204 (w), 1166 (w), 1146 (m), 1129 (w), 1089 (w), 1013 (w), 922 (w), 894 (w), 856 (w), 827 (vs), 802 (w), 749 (w), 731 (w), 704 (m), 691 (s), 588 (s), 550 (w), 514 (w), 497 (w) cm⁻¹. – **MS** (EI, 70 eV), *m/z* (%): 552 [M+H]⁺, 551 [M]⁺. – **HRMS** (FAB, 3-NBA) calc. for C₃₇H₄₇N₃F₁ [M+H]⁺ 552.3754, found 552.3755.

2,4-Bis(4-(*tert*-butyl)phenyl)-6-(3,4,5-trifluorophenyl)-1,3,5-triazine (95)

Under argon atmosphere, a mixture of 2,4-bis(4-tert-butylphenyl)-6-chloro-1,3,5-triazine (1.14 g, 3.00 mmol, 1.00 equiv.), (3,4,5-trifluorophenyl)boronic acid (554 mg, 3.15 mmol, 1.05 equiv.), tetrakis(triphenylphosphine)palladium(0) (172 mg, 149 μmol, 5 mol%) and aqueous potassium carbonate (4M, 4 mL, 16.00 mmol, 5.33 equiv.) in PhMe/EtOH (40/4 mL) was stirred at 90 °C for 12 h. After cooling to room temperature, the reaction mixture was diluted with 50 mL of ethyl acetate and washed with brine (3 × 50 mL). The organic layer was dried over MgSO$_4$ and the solvent was removed under reduced pressure. The obtained crude product was purified *via* column chromatography on silica gel (cyclohexane/dichloromethane = 15:1 to 10:1) to yield the title compound as a white solid (1.35 g, 2.85 mmol, 95%).

R_f = 0.30 (cyclohexane/ethyl acetate = 50:1). – **^{1}H NMR** (400 MHz, CDCl$_3$) δ = 8.61 (d, J = 8.5 Hz, 4H), 8.36 (dd, J = 8.6 Hz, J = 6.8 Hz, 2H), 7.59 (d, J = 8.5 Hz, 4H), 1.41 (s, 18H) ppm. – **^{13}C NMR** (101 MHz, CDCl$_3$) δ = 171.9 (C$_q$, C$_{Ar}$), 156.7 (C$_q$, C$_{Ar}$), 133.1 (C$_q$, C$_{Ar}$), 129.0 (+, C$_{Ar}$H), 125.8 (+, C$_{Ar}$H), 113.3–113.0 (+, m, C$_{Ar}$H), 35.3 (C$_q$, C(CH$_3$)$_3$), 31.3 (+, CH$_3$) ppm. – **^{19}F NMR** (376 MHz, CDCl$_3$) δ = –137.83 (dd, J = 20.4 Hz, J = 8.6 Hz, 2F), –158.77 (tt, J = 20.4 Hz, J = 6.8 Hz, 1F) ppm. – **IR** (ATR, ṽ) = 2953 (w), 2902 (w), 2867 (vw), 1611 (w), 1575 (w), 1530 (vs), 1509 (vs), 1463 (w), 1435 (w), 1418 (s), 1411 (m), 1368 (vs), 1319 (w), 1309 (w), 1268 (w), 1227 (m), 1191 (w), 1105 (m), 1045 (s), 1016 (m), 993 (vw), 887 (w), 858 (w), 813 (vs), 738 (m), 698 (vs), 577 (s), 565 (w), 544 (m), 514 (vw) cm^{-1}. – **MS** (EI, 70 eV), *m/z* (%): 476 [M+H]$^+$, 475 [M]$^+$. – **HRMS** (FAB, 3-NBA) calc. for C$_{29}$H$_{29}$N$_3$F$_3$ [M+H]$^+$ 476.2314, found 476.2313.

5,11-Dihydroindolo[3,2-b]carbazole (76)

10 mL trimethyl orthoformate and 1 mL of concentrated H_2SO_4 was added to a stirred solution of 3-(1H-indol-3-ylmethyl)-1H-indole (15.0 g, 60.9 mmol, 1.00 equiv.) in methanol (200 mL). The reaction mixture was refluxed at 70 °C for 1 h. The resulting product mixture was filtered and washed with cold methanol to yield the title compound as a dark brown solid (10.00 g, 39 mmol, 64%).

R_f = 0.50 (dichloromethane = 1). – **¹H NMR** (400 MHz, DMSO-d_6) δ = 11.02 (s, 2H), 8.19 (d, J = 7.7 Hz, 2H), 8.11 (s, 2H), 7.46 (d, J = 8.0 Hz, 2H), 7.37 (t, J = 7.6 Hz, 2H), 7.12 (t, J = 7.8 Hz, 2H) ppm. – **¹³C NMR** (101 MHz, DMSO-d_6) δ = 141.1 (C_q, 2C_{Ar}), 135.0 (C_q, 2C_{Ar}), 125.4 (+, 2C_{Ar}H), 122.6 (C_q, 2C_{Ar}), 122.6 (C_q, 2C_{Ar}), 120.2 (+, 2C_{Ar}H), 117.6 (+, 2C_{Ar}H), 110.5 (+, 2C_{Ar}H), 100.5 (+, 2C_{Ar}H) ppm. – **IR** (ATR, ṽ) = 3395 (s), 1613 (w), 1521 (w), 1456 (m), 1446 (s), 1320 (m), 1266 (w), 1232 (s), 1179 (m), 1142 (w), 1000 (w), 856 (m), 847 (s), 739 (vs), 688 (vs), 568 (m), 426 (vs), 411 (vs), 385 (vs) cm^{-1}. – **MS** (EI, 70 eV), m/z: 256 [M]$^+$. – **HRMS** (EI, 70 eV) calc. for $C_{18}H_{12}N_2$ [M]$^+$ 256.1000, found 256.1001. The analytical data is consistent with literature.[136]

5,11-Bis(4-(4,6-bis(4-(tert-butyl)phenyl)-1,3,5-triazin-2-yl)phenyl)-5,11-dihydroindolo[3,2-b]carbazole (85, ICzTRZ)

A 50 mL sealed vial was charged with 5,11-dihydroindolo[3,2-b]carbazole (250 mg, 975 μmol, 1.00 equiv.), 2,4-bis(4-(tert-butyl)phenyl)-6-(4-fluorophenyl)-1,3,5-triazine (900 mg, 2.05 mmol, 2.10 equiv.) and potassium phosphate tribasic (2.07 g, 9.75 mmol, 10.00 equiv.). It was evacuated and flushed with argon three times. Through the septum 20 mL anhydrous methyl sulfoxide was added, then it was heated to 120 °C and stirred for 12 h. The reaction mixture was diluted with dichloromethane (50 mL) and washed with brine (3 × 50 mL). The organic layer was dried over MgSO$_4$ and the solvent was removed under reduced pressure. The crude product was purified by column chromatography on silica gel (cyclohexane/dichloromethane = 10:1 to 5:1) to yield the title compound as a yellow luminescent solid (700 mg, 639 μmol, 66%).

R_f = 0.30 (cyclohexane/dichloromethane = 5:1). – ^{1}H NMR (400 MHz, CDCl$_3$) δ = 9.09 (d, J = 8.5 Hz, 4H), 8.76 (d, J = 8.5 Hz, 8H), 8.25 (s, 2H), 8.18 (d, J = 7.8 Hz, 2H), 7.95 (d, J = 8.5 Hz, 4H), 7.65 (d, J = 8.6 Hz, 8H), 7.61 (d, J = 8.2 Hz, 2H), 7.47 (t, J = 8.1 Hz, 2H), 7.31 (t, J = 7.4 Hz, 2H), 1.44 (s, 36H) ppm. – ^{13}C NMR (100 MHz, CDCl$_3$) δ = 171.8 (C$_q$, 4C$_{Ar}$), 170.9 (C$_q$, 2C$_{Ar}$), 156.3 (C$_q$, 4C$_{Ar}$), 142.2 (C$_q$, 2C$_{Ar}$), 141.7 (C$_q$, 2C$_{Ar}$), 136.9 (C$_q$, 2C$_{Ar}$), 135.2 (C$_q$, 2C$_{Ar}$), 133.7 (C$_q$, 4C$_{Ar}$), 130.9 (+, 4C$_{Ar}$H), 129.0 (+, 8C$_{Ar}$H), 126.9 (+, 4C$_{Ar}$H), 126.5 (+, 2C$_{Ar}$H), 125.8 (+, 8C$_{Ar}$H), 124.1 (C$_q$, 2C$_{Ar}$), 124.0 (C$_q$, 2C$_{Ar}$), 120.6 (+, 2C$_{Ar}$H), 120.1 (+, 2C$_{Ar}$H), 109.9 (+, 2C$_{Ar}$H), 100.4 (+, 2C$_{Ar}$H), 35.3 (C$_q$, 4C(CH$_3$)$_3$), 31.4 (+, 12CH$_3$) ppm. – **IR** (ATR, ṽ) = 2958 (w), 1604 (w), 1578 (m), 1504 (vs), 1477 (s), 1460 (m), 1441 (vs), 1408 (s), 1366 (vs), 1324 (m), 1303 (m), 1266 (w), 1234 (m), 1190 (m), 1169 (w), 1149 (w), 1106 (w), 1016 (w), 847 (w), 815 (vs), 758 (w), 738 (s), 701 (w), 565 (m), 550 (m) cm^{-1}. – **HRMS** (ESI) calc. for C$_{76}$H$_{70}$N$_8$ [M]$^+$ 1094.5723, found 1094.5715. – **EA** (C$_{76}$H$_{70}$N$_8$) calc. C: 83.33, H: 6.44, N: 10.23; found C: 83.32, H: 6.50, N: 10.12.

5,11-Bis(4-(4,6-bis(4-(tert-butyl)phenyl)-1,3,5-triazin-2-yl)-2,6-difluorophenyl)-5,11-dihydroindolo[3,2-b]carbazole (86)

A 50 mL sealable vial was charged with 5,11-dihydroindolo[3,2-b]carbazole (192 mg, 749 μmol, 1.00 equiv.), 2,4-bis(4-(tert-butyl)phenyl)-6-(3,4,5-trifluorophenyl)-1,3,5-triazine (749 mg, 1.58 mmol, 2.10 equiv.) and potassium phosphate tribasic (1.59 g, 7.51 mmol, 10.00 equiv.). It was evacuated and flushed with argon three times. Through the septum 15 mL of anhydrous DMSO were added, then it was heated to 120 °C and stirred for 12 h. After cooling to room temperature, the reaction mixture was diluted with 100 mL of dichloromethane and washed with brine (3 × 50 mL). The organic layer was dried over MgSO$_4$ and the solvent was removed under reduced pressure. The obtained crude product was purified *via* column chromatography on silica gel (cyclohexane/dichloromethane = 5:1 to 2:1) to yield the title compound as a yellow luminescent solid (500 mg, 428 μmol, 57%).

R$_f$ = 0.45 (cyclohexane/dichloromethane = 3:1). – **^{1}H NMR** (400 MHz, CDCl$_3$) δ = 8.75 (d, J = 8.5 Hz, 8H), 8.69 (d, J = 8.7 Hz, 4H), 8.17 (d, J = 7.8 Hz, 2H), 7.91 (d, J = 1.5 Hz, 2H), 7.65 (d, J = 8.5 Hz, 8H), 7.48 (t, J = 7.6 Hz, 2H), 7.32 (t, J = 7.5 Hz, 2H), 7.29–7.22 (m, 2H), 1.44 (s, 36H) ppm. – **^{19}F NMR** (376 MHz, CDCl$_3$) δ = –117.84 ppm. – **IR** (ATR, ṽ) = 2963 (w), 2905 (w), 1609 (w), 1578 (w), 1506 (vs), 1477 (s), 1445 (vs), 1409 (m), 1367 (vs), 1360 (vs), 1339 (m), 1322 (m), 1265 (m), 1231 (s), 1196 (m), 1170 (w), 1150 (w), 1105 (m), 1034 (s), 1018 (w), 888 (w), 857 (w), 834 (w), 815 (vs), 764 (w), 755 (w), 735 (vs), 720 (w), 708 (m), 681 (w), 581 (m), 564 (m), 547 (w), 535 (w), 517 (w), 429 (w), 411 (w) cm^{-1}. – **HRMS** (ESI) calc. for C$_{76}$H$_{66}$N$_8$F$_4$ [M]$^+$ 1166.5347, found 1166.5344.

5,11-Bis(4-(4,6-bis(3,5-di-tert-butylphenyl)-1,3,5-triazin-2-yl)phenyl)-5,11 dihydroindolo[3,2-b]carbazole (87)

A 50 mL sealable vial was charged with 5,11-dihydroindolo[3,2-b]carbazole (85.0 mg, 332 μmol, 1.00 equiv.), 2,4-bis(3,5-di-tert-butylphenyl)-6-(4-fluorophenyl)-1,3,5-triazine (386 mg, 700 μmol, 2.11 equiv.) and potassium phosphate tribasic (707 mg, 3.33 mmol, 10.00 equiv.). It was evacuated and flushed with argon three times. Through the septum 15 mL of anhydrous DMSO were added, then it was heated to 120 °C and stirred for 12 h. After cooling to room temperature, the reaction mixture was diluted with 100 mL of dichloromethane and washed with brine (3 × 50 mL). The organic layer was dried over MgSO$_4$ and the solvent was removed under reduced pressure. The obtained crude product was purified *via* column chromatography on silica gel (cyclohexane/dichloromethane = 10:1 to 5:1) to yield the title compound as a luminescent yellow solid (300 mg, 227 μmol, 69%).

R$_f$ = 0.50 (cyclohexane/ethyl acetate = 50:1). – **^{1}H NMR** (400 MHz, CDCl$_3$) δ = 9.12 (d, J = 8.5 Hz, 4H), 8.75 (d, J = 1.9 Hz, 8H), 8.31 (s, 2H), 8.21 (dd, J = 7.8 Hz, J = 1.1 Hz, 2H), 8.01 (d, J = 8.6 Hz, 4H), 7.75 (t, J = 1.9 Hz, 4H), 7.64 (d, J = 8.2 Hz, 2H), 7.48 (t, J = 7.7 Hz, 2H), 7.32 (t, J = 7.4 Hz, 2H), 1.50 (s, 72H) ppm. – **^{13}C NMR** (101 MHz, CDCl$_3$) δ = 172.4 (C$_q$, C$_{Ar}$), 171.0 (C$_q$, C$_{Ar}$), 151.4 (C$_q$, C$_{Ar}$), 142.2 (C$_q$, C$_{Ar}$), 141.8 (C$_q$, C$_{Ar}$), 136.9 (C$_q$, C$_{Ar}$), 135.7 (C$_q$, C$_{Ar}$), 135.2 (C$_q$, C$_{Ar}$), 130.9 (+, C$_{Ar}$H), 127.1 (+, C$_{Ar}$H), 127.0 (+, C$_{Ar}$H), 126.5 (+, C$_{Ar}$H), 124.1 (C$_q$, C$_{Ar}$), 124.0 (C$_q$, C$_{Ar}$), 123.4 (+, C$_{Ar}$H), 120.6 (+, C$_{Ar}$H), 120.2 (+, C$_{Ar}$H), 110.0 (+, C$_{Ar}$H), 100.5 (+, C$_{Ar}$H), 35.3 (C$_q$, *C*(CH$_3$)$_3$), 31.7 (+, *C*H$_3$) ppm. – **IR** (ATR, ṽ) = 2953 (w), 2904 (w), 2866 (w), 1605 (w), 1516 (vs), 1477 (w), 1451 (s), 1442 (s), 1408 (w), 1392 (w), 1363 (s), 1354 (vs), 1320 (w), 1296 (w), 1283 (w), 1266 (w), 1247 (w), 1235 (m), 1203 (w), 1188 (w), 1166 (w), 1156 (w), 1020 (w), 894 (w), 837 (w), 827 (s), 738 (s), 707 (w), 691 (m), 562 (w) cm^{-1}. – **HRMS** (ESI) calc. for C$_{92}$H$_{102}$N$_8$ [M]$^+$ 1318.8227, found 1318.8224.

5,11-Bis(4-(4,6-diphenyl-1,3,5-triazin-2-yl)phenyl)-5,11-dihydroindolo[3,2-b]carbazole (96)

A 50 mL sealable vial was charged with 5,11-dihydroindolo[3,2-b]carbazole (192 mg, 749 µmol, 1.00 equiv.), 2-(4-fluorophenyl)-4,6-diphenyl-1,3,5-triazine (515 mg, 1.57 mmol, 2.10 equiv.) and potassium phosphate tribasic (1.59 g, 7.50 mmol, 10.00 equiv.). It was evacuated and flushed with argon three times. Through the septum 20 mL of anhydrous DMSO were added, then it was heated to 120 °C and stirred for 12 h. After cooling to room temperature, 100 mL water was added to the mixture. Then it was filtered and the solid was thoroughly washed with methanol and dichloromethane. The remaining yellow solid was the title compound as a luminescent yellow solid (520 mg, 597 µmol, 80%).

NMR measurements were not successful due to low solubility.

– **IR** (ATR, ṽ) = 2921 (w), 2853 (w), 1738 (w), 1599 (m), 1588 (m), 1572 (w), 1510 (vs), 1475 (s), 1439 (vs), 1412 (s), 1361 (vs), 1317 (vs), 1303 (s), 1293 (s), 1228 (s), 1198 (m), 1190 (m), 1173 (s), 1164 (s), 1142 (s), 1106 (m), 1064 (m), 1027 (m), 1014 (m), 1000 (w), 969 (w), 952 (w), 929 (m), 851 (w), 840 (w), 829 (s), 819 (w), 766 (vs), 755 (w), 737 (s), 722 (vs), 686 (vs), 676 (s), 662 (m), 645 (m), 637 (m), 629 (w), 564 (s), 509 (m), 466 (w), 408 (w) cm^{-1}. – **HRMS** (ESI) calc. for $C_{60}H_{38}N_8$ [M]$^+$ 870.3219, found 870.3213. – **EA** ($C_{60}H_{38}N_8$) calc. C: 82.74, H: 4.40, N: 12.86; found C: 80.53, H: 4.35, N: 12.38.

9-(4-(4,6-Bis(4-(tert-butyl)phenyl)-1,3,5-triazin-2-yl)phenyl)-9H-carbazole (106)

A 50 mL sealable vial was charged with (4-carbazol-9-ylphenyl)boronic acid (431 mg, 1.50 mmol, 1.00 equiv.), 2,4-bis(4-tert-butylphenyl)-6-chloro-1,3,5-triazine (598 mg, 1.57 mmol, 1.05 equiv.), Pd(PPh$_3$)$_4$ (43.0 mg, 37.2 µmol, 2.5 mol%). It was evacuated and flushed with argon three times. Through the septum 2.0 mL aqueous potassium carbonate (4M, 2 mL, 8.00 mmol, 5.33 equiv.), 10 mL toluene and 2.0 mL ethanol were added, then it was heated to 90 °C and stirred for 2.5 h. After cooling to room temperature, the reaction mixture was diluted with 30 mL of ethyl acetate and washed with brine (3 × 30 mL). The organic layer was dried over MgSO$_4$ and the solvent was removed under reduced pressure. The obtained crude product was purified *via* column chromatography on silica gel (cyclohexane/dichloromethane = 10:1) to yield the title compound as a white luminescent solid (454 mg, 774 µmol, 52%).

R$_f$ = 0.55 (cyclohexane/dichloromethane = 10:1). – **^{1}H NMR** (400 MHz, CDCl$_3$) δ = 9.02 (d, J = 8.6 Hz, 2H), 8.74 (d, J = 8.6 Hz, 4H), 8.18 (d, J = 7.6 Hz, 2H), 7.82 (d, J = 8.6 Hz, 2H), 7.63 (d, J = 8.6 Hz, 4H), 7.58 (d, J = 8.2 Hz, 2H), 7.47 (t, J = 8.2 Hz, 2H), 7.34 (t, J = 7.4 Hz, 2H), 1.43 (s, 18H) ppm. – **^{13}C NMR** (101 MHz, CDCl$_3$) δ = 171.8 (C$_q$, C$_{Ar}$), 170.8 (C$_q$, C$_{Ar}$), 156.3 (C$_q$, C$_{Ar}$), 141.6 (C$_q$, C$_{Ar}$), 140.6 (C$_q$, C$_{Ar}$), 135.4 (C$_q$, C$_{Ar}$), 133.7 (C$_q$, C$_{Ar}$), 130.7 (+, C$_{Ar}$H), 129.0 (+, C$_{Ar}$H), 126.8 (+, C$_{Ar}$H), 126.3 (+, C$_{Ar}$H), 125.8 (+, C$_{Ar}$H), 123.9 (C$_q$, C$_{Ar}$), 120.6 (+, C$_{Ar}$H), 120.5 (+, C$_{Ar}$H), 110.1 (+, C$_{Ar}$H), 35.3 (C$_q$, CCH$_3$), 31.4 (+, CH$_3$) ppm. – **IR** (ATR, ṽ) = 2959 (w), 1605 (w), 1577 (m), 1503 (vs), 1477 (s), 1448 (vs), 1409 (s), 1366 (vs), 1333 (s), 1313 (s), 1265 (m), 1225 (s), 1190 (m), 1170 (m), 1150 (m), 1105 (m), 1016 (m), 850 (w), 815 (vs), 747 (vs), 722 (vs), 713 (m), 704 (m), 622 (w), 550 (s) cm^{-1}. – **MS** (EI, 70 eV), m/z (%): 587 [M+H]$^+$, 586 [M]$^+$. – **HRMS** (FAB, 3-NBA) calc. for C$_{41}$H$_{39}$N$_4$ [M+H]$^+$ 587.3175, found 587.3173. – **EA** (C$_{41}$H$_{38}$N$_4$) calc. C: 83.92, H: 6.53, N: 9.55; found C: 83.77, H: 6.37, N: 9.46.

9,9'-Bis(4-(4,6-bis(4-(tert-butyl)phenyl)-1,3,5-triazin-2-yl)phenyl)-9H,9'H-3,3'-bicarbazole
(103, DCzTRZ)

A 100 mL round flask was charged with $FeCl_3$ (162 mg, 1000 µmol, 4.00 equiv.) and DCM (8.0 mL) under argon. Then, 9-(4-(4,6-bis(4-(tert-butyl)phenyl)-1,3,5-triazin-2-yl)phenyl)-9H-carbazole (146.7 mg, 250 µmol, 1.00 equiv.) in DCM (3.0 mL) was very slowly added. The mixture was stirred at room temperature for 5 h. After 10% sodium hydroxide solution was added into the mixture, the aqueous solution was thoroughly extracted with DCM. The organic layer was dried over $MgSO_4$ and the solvent was removed under reduced pressure. The obtained crude product was purified *via* column chromatography on silica gel (cyclohexane/dichloromethane = 5:1) to yield the title compound as a white luminescent solid (85 mg, 72.6 µmol, 58%).

R_f = 0.60 (cyclohexane/dichloromethane = 5:1). – ^{1}H NMR (400 MHz, CDCl$_3$) δ = 9.04 (d, J = 8.6 Hz, 4H), 8.74 (d, J = 8.6 Hz, 8H), 8.51 (d, J = 1.5 Hz, 2H), 8.29 (d, J = 7.7 Hz, 2H), 7.91–7.79 (m, 6H), 7.72–7.56 (m, 12H), 7.50 (t, J = 7.7 Hz, 2H), 7.38 (t, J = 7.5 Hz, 2H), 1.43 (s, 36H) ppm. – ^{13}C NMR (101 MHz, CDCl$_3$) δ = 171.8 (C$_q$, C$_{Ar}$), 170.8 (C$_q$, C$_{Ar}$), 156.3 (C$_q$, C$_{Ar}$), 141.6 (C$_q$, C$_{Ar}$), 141.1 (C$_q$, C$_{Ar}$), 139.8 (C$_q$, C$_{Ar}$), 135.4 (C$_q$, C$_{Ar}$), 134.8 (C$_q$, C$_{Ar}$), 133.7 (C$_q$, C$_{Ar}$), 130.8 (+, C$_{Ar}$H), 129.0 (+, C$_{Ar}$H), 126.7 (+, C$_{Ar}$H), 126.4 (+, C$_{Ar}$H), 126.1 (+, C$_{Ar}$H), 125.8 (+, C$_{Ar}$H), 124.5 (C$_q$, C$_{Ar}$), 124.1 (C$_q$, C$_{Ar}$), 120.7 (+, C$_{Ar}$H), 120.6 (+, C$_{Ar}$H), 119.1 (+, C$_{Ar}$H), 110.4 (+, C$_{Ar}$H), 110.2 (+, C$_{Ar}$H), 35.3 (C$_q$, CCH$_3$), 31.4 (+, CH$_3$) ppm. – IR (ATR, $\tilde{v}$) = 2959 (w), 1605 (w), 1578 (w), 1503 (vs), 1469 (s), 1448 (vs), 1408 (s), 1366 (vs), 1357 (vs), 1324 (s), 1265 (m), 1225 (m), 1190 (m), 1170 (m), 1152 (w), 1106 (w), 1016 (w), 815 (vs), 799 (m), 742 (m), 727 (m), 704 (w), 550 (m) cm^{-1}. – HRMS (ESI) calc. for $C_{82}H_{74}N_8$ [M]$^+$ 1170.6036, found 1170.6035. – EA ($C_{82}H_{74}N_8$) calc. C: 84.07, H: 6.37, N: 9.56; found C: 82.81, H: 6.70, N: 8.93.

5,5',11,11'-Tetrakis(4-(4,6-bis(4-(tert-butyl)phenyl)-1,3,5-triazin-2-yl)phenyl)-5,5',11,11'-tetrahydro-2,2'-biindolo[3,2-b]carbazole (104, DICzTRZ)

A 50 mL round flask was charged with FeCl$_3$ (32.4 mg, 200 µmol, 3.73 equiv.) and DCM (4.0 mL) under Argon. Then, 9,9'-bis(4-(4,6-bis(4-(tert-butyl)phenyl)-1,3,5-triazin-2-yl)phenyl)-9H,9'H-3,3'-bicarbazole (58.5 mg, 53.4 µmol, 1.00 equiv.) in DCM (2.0 mL) was very slowly added. The mixture was stirred at 60 °C for 12 h. After 10% sodium hydroxide solution was added into the mixture, the aqueous solution was thoroughly extracted with DCM. The organic layer was dried over MgSO$_4$ and the solvent was removed under reduced pressure. The obtained crude product was purified *via* column chromatography on silica gel (cyclohexane/dichloromethane = 2:1) to yield the title compound as a yellow luminescent solid (38.5 mg, 17.6 µmol, 66%).

R$_f$ = 0.40 (cyclohexane/dichloromethane = 2:1). – **^{1}H NMR** (500 MHz, CDCl$_3$) δ = 9.07 (d, *J* = 7.8 Hz, 8H), 8.77 (d, *J* = 8.3 Hz, 8H), 8.67 (d, *J* = 7.5 Hz, 8H), 8.46 (s, 2H), 8.31 (s, 2H), 8.19 (s, 2H), 8.15 (d, *J* = 7.1 Hz, 2H), 7.91 (d, *J* = 6.9 Hz, 8H), 7.83 (d, *J* = 8.2 Hz, 2H), 7.67–7.45 (m, 20H), 7.46 (d, *J* = 7.3 Hz, 2H), 7.32 (d, *J* = 7.3 Hz, 2H), 1.47 (s, 36H), 1.34 (s, 36H) ppm. – **^{13}C NMR** (125 MHz, CDCl$_3$) δ = 171.6, 171.5, 170.7, 170.6, 156.1, 156.0, 142.1, 142.0, 141.4, 140.5, 137.0, 136.5, 134.7, 134.5, 133.6, 133.5, 130.7, 128.9, 128.8, 126.4, 126.2, 125.7, 125.6, 124.6, 124.0, 123.9, 120.5, 120.0, 119.2, 113.9, 110.0, 109.8, 100.4, 100.3, 35.2, 35.0, 31.3, 31.2 ppm. – **IR** (ATR, ṽ) = 2959 (w), 2932 (w), 2904 (w), 1605 (w), 1578 (m), 1503 (vs), 1476 (s), 1446 (vs), 1408 (s), 1367 (vs), 1356 (vs), 1322 (m), 1312 (m), 1264 (m), 1231 (m), 1190 (m), 1173 (m), 1150 (m), 1128 (w), 1106 (m), 1057 (w), 1016 (m), 928 (w), 856 (w), 849 (w), 833 (w), 815 (vs), 795 (m), 762 (w), 735 (m), 727 (w), 704 (m), 684 (w), 602 (w), 577 (m), 550 (s), 509 (w), 480 (w), 453 (w), 414 (w), 377 (w) cm^{-1}. – **HRMS** (ESI) calc. for C$_{152}$H$_{138}$N$_{16}$ [M]$^+$ 2188.1324, found 2188.1383.

2, 2', 2'', 5'-Tetrabromo-1,1':4',1''-terphenyl (127)

Under argon atmosphere, a mixture of 1,4-dibromo-2,5-diiodobenzene (2.13 g, 4.37 mmol, 1.00 equiv.), (2-bromophenyl)boronic acid (1.80 g, 8.96 mmol, 2.05 equiv.), Ag_2CO_3 (6.00 g, 21.8 mmol, 4.98 equiv.) and Pd(PPh$_3$)$_4$ (500 mg, 433 µmol, 10 mol%) in THF/H$_2$O (50/25 mL) was stirred at 80 °C for about 12 h. TLC was used to monitor the progress of reaction. After full conversion, the mixture was cooled to room temperature and diluted with ethyl acetate (50 mL). Then it was washed with brine (2 × 50 mL), dried over MgSO$_4$ and the solvent was removed under reduced pressure. The crude product was purified by column chromatography on silica gel (cyclohexane/ethyl acetate = 100:1) to yield the target compound 1,4-dibromo-2,5-bis(2-bromophenyl)benzene (1.86 g, 3.41 mmol, 78%) as a colorless solid.

R_f = 0.50 (cyclohexane/ethyl acetate = 100:1). – ^{1}H NMR (400 MHz, CDCl$_3$) δ = 7.70 (m, 2H), 7.56 (d, J = 2.0 Hz, 2H), 7.39–7.44 (m, 2H), 7.35–7.44 (m, 4H) ppm. – ^{13}C NMR (101 MHz, CDCl$_3$) δ = 143.1 (C$_q$, C$_{Ar}$), 143.1 (C$_q$, C$_{Ar}$), 140.7 (C$_q$, C$_{Ar}$), 140.6 (C$_q$, C$_{Ar}$), 134.6 (+, C$_{Ar}$H), 134.6 (+, C$_{Ar}$H), 132.9 (+, C$_{Ar}$H), 132.8 (+, C$_{Ar}$H), 131.2 (+, C$_{Ar}$H), 130.9 (+, C$_{Ar}$H), 130.0 (+, C$_{Ar}$H), 130.0 (+, C$_{Ar}$H), 127.4 (+, C$_{Ar}$H), 127.4 (+, C$_{Ar}$H), 123.5 (C$_q$, C$_{Ar}$), 123.4 (C$_q$, C$_{Ar}$), 122.3 (C$_q$, C$_{Ar}$), 122.2 (C$_q$, C$_{Ar}$) ppm. – IR (ATR, ṽ) = 2925, 1561, 1445, 1428, 1421, 1344, 1159, 1091, 1052, 1027, 1004, 984, 948, 887, 755, 730, 693, 659, 647, 557, 482, 460, 443 cm^{-1}. – MS (EI, 70 eV), m/z (%): 545 [M]$^+$. – HRMS (EI, C$_{18}$H$_{10}$^{79}Br$_4$) calc.: 541.7516; found: 541.7519. HRMS (EI, C$_{18}$H$_{10}$^{79}Br$_2$^{81}Br$_2$), calc.: 545.7475; found, 545.7478. The analytical data is consistent with literature.[143]

5,11-Bis(4-chloro-2,6-dimethylphenyl)-5,11-dihydroindolo[3,2-b]carbazole (125)

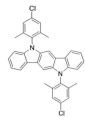

Under argon atmosphere, a mixture of 1,4-dibromo-2,5-bis(2-bromophenyl)benzene (1.05 g, 1.92 mmol, 1.00 equiv.), 4-chloro-2,6-dimethylaniline (637 mg, 4.09 mmol, 2.13 equiv.), Pd$_2$(dba)$_3$ (90.0 mg, 98.3 μmol, 5 mol%), ditert-butyl(phenyl)phosphane (83.0 mg, 373 μmol, 19 mol%) and NaO^{t}Bu (1.13 g, 11.7 mmol, 6.09 equiv.) in PhMe/^{t}BuOH (14/2 mL) was stirred at 100 °C for 12 h. TLC was used to monitor the progress of reaction. After full conversion, the mixture was cooled to room temperature and diluted with DCM (50 mL). Then it was washed with brine (2 × 50 mL), dried over MgSO$_4$ and the solvent was removed under reduced pressure. The crude product was purified by column chromatography on silica gel (cyclohexane/dichloromethane = 10:1 to 3:1) to yield the target compound 5,11-bis(4-chloro-2,6-dimethylphenyl)-5,11-dihydroindolo[3,2-*b*]carbazole (760 mg, 1.42 mmol, 74%) as a colorless solid.

R$_f$ = 0.70 (cyclohexane/ethyl acetate = 50:1). – ^{1}H NMR (400 MHz, CDCl$_3$) δ = 8.11 (d, *J* = 7.5 Hz, 2H), 7.58 (s, 2H), 7.38 (ddd, *J* = 8.3 Hz, *J* = 7.2 Hz, *J* = 1.3 Hz, 2H), 7.36 (bs, 4H), 7.25–7.21 (m, 6H), 6.93 (d, *J* = 8.1 Hz, 2H), 1.92 (s, 12H) ppm. – ^{13}C NMR (101 MHz, CDCl$_3$) δ = 141.3 (C$_q$, 2C$_{Ar}$), 140.6 (C$_q$, 4C$_{Ar}$), 136.3 (C$_q$, 2C$_{Ar}$), 134.3 (C$_q$, 2C$_{Ar}$), 133.8 (C$_q$, 2C$_{Ar}$), 128.9 (+, 4C$_{Ar}$H), 126.4 (+, 2C$_{Ar}$H), 123.3 (C$_q$, 2C$_{Ar}$), 123.1 (C$_q$, 2C$_{Ar}$), 120.6 (+, 2C$_{Ar}$H), 119.2 (+, 2C$_{Ar}$H), 109.1 (+, 2C$_{Ar}$H), 99.7 (+, 2C$_{Ar}$H), 17.8 (+, 4CH$_3$) ppm. – IR (ATR, ṽ) = 2921, 1611, 1581, 1504, 1473, 1452, 1441, 1378, 1363, 1336, 1319, 1292, 1261, 1231, 1207, 1186, 1164, 1145, 1108, 1031, 1020, 997, 933, 924, 850, 764, 749, 744, 720, 690, 581, 568, 533, 477, 436, 416, 381 cm^{-1}. – MS (FAB, 3-NBA), *m/z* (%): 532 [M]$^+$. – HRMS (FAB, 3-NBA, C$_{34}$H$_{26}$N$_2$^{35}Cl$_2$) calc.: 532.1473; found: 532.1474.

5,11-Bis(2,6-dimethyl-4-(4,4,5,5-tetramethyl-1,3,2-dioxaborolan-2-yl)phenyl)-5,11-dihydroindolo[3,2-b]carbazole (128)

Under argon atmosphere, a mixture of 5,11-bis(4-chloro-2,6-dimethylphenyl)-5,11-dihydroindolo[3,2-b]carbazole (2.12 g, 3.97 mmol, 1.00 equiv.), 4,4,5,5-tetramethyl-2-(4,4,5,5-tetramethyl-1,3,2-dioxaborolan-2-yl)-1,3,2-dioxaborolane (6.08 g, 23.9 mmol, 6.03 equiv.), Pd$_2$(dba)$_3$ (368 mg, 402 μmol, 10 mol%), XPhos (380 mg, 797 μmol, 20 mol%) and KOAc (4.68 g, 47.7 mmol, 12.0 equiv.) in 40 mL 1,4-dioxane was stirred at 110 °C for 12 h. TLC was used to monitor the progress of reaction. After full conversion, the mixture was cooled to room temperature and diluted with DCM (150 mL). Then it was washed with brine (2 × 150 mL), dried over MgSO$_4$ and the solvent was removed under reduced pressure. The crude product was purified by column chromatography on silica gel (cyclohexane/dichloromethane = 2:1 to 1:1) to yield the target compound 5,11-bis(2,6-dimethyl-4-(4,4,5,5-tetramethyl-1,3,2-dioxaborolan-2-yl)phenyl)-5,11-dihydroindolo[3,2-b]carbazole (2.79 g, 3.89 mmol, 98%) as a colorless solid.

R$_f$ = 0.65 (cyclohexane/ethyl acetate = 10:1). – ^{1}H NMR (400 MHz, CDCl$_3$) δ = 8.08 (d, J = 7.7 Hz, 2H), 7.81 (s, 4H), 7.58 (s, 2H), 7.35 (t, J = 7.6 Hz, 2H), 7.20 (t, J = 7.2 Hz, 2H), 6.90 (d, J = 8.1 Hz, 2H), 1.95 (s, 12H), 1.43 (s, 24H) ppm. – ^{13}C NMR (101 MHz, CDCl$_3$) δ = 141.3 (C$_q$, 2C$_{Ar}$), 138.1 (C$_q$, 2C$_{Ar}$), 138.0 (C$_q$, 2C$_{Ar}$), 136.3 (C$_q$, 2C$_{Ar}$), 135.3 (+, 4C$_{Ar}$H), 126.1 (+, 2C$_{Ar}$H), 123.3 (C$_q$, 2C$_{Ar}$), 123.1 (C$_q$, 2C$_{Ar}$), 120.5 (+, 2C$_{Ar}$H), 118.9 (+, 2C$_{Ar}$H), 109.2 (+, 2C$_{Ar}$H), 99.8 (+, 2C$_{Ar}$H), 84.3 (C$_q$, 4C$_{Ar}$), 29.9 (C$_q$, 4COCCH$_3$CH$_3$), 25.1 (+, 8CH$_3$COC), 17.7 (+, 4CH$_3$C$_{Ar}$) ppm. – IR (ATR, ṽ) = 2980, 1604, 1504, 1475, 1453, 1442, 1404, 1385, 1373, 1357, 1337, 1310, 1295, 1275, 1232, 1190, 1166, 1143, 1115, 1004, 992, 967, 941, 928, 881, 851, 833, 761, 742, 705, 687, 670, 579, 562, 518, 438, 416 cm^{-1}. – MS (FAB, 3-NBA), m/z (%): 716 [M]$^+$. – HRMS (FAB, 3-NBA, C$_{46}$H$_{50}$B$_2$N$_2$O$_4$) calc.: 716.3957; found: 716.3956.

5,11-Bis(3,5-dimethyl-[1,1'-biphenyl]-4-yl)-5,11-dihydroindolo[3,2-b]carbazole (108)

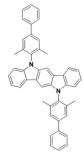

Under argon atmosphere, a mixture of 5,11-bis(2,6-dimethyl-4-(4,4,5,5-tetramethyl-1,3,2-dioxaborolan-2-yl)phenyl)-5,11-dihydroindolo[3,2-b]carbazole (179 mg, 250 µmol, 1.00 equiv.), bromobenzene (98.0 mg, 624 µmol, 2.50 equiv.), Pd(PPh$_3$)$_4$ (29.0 mg, 25.1 µmol, 10 mol%) and K$_2$CO$_3$ (345 mg, 2.50 mmol, 9.99 equiv.) in PhMe/EtOH/H$_2$O (8.00/1.25/1.25 mL) was stirred at 90 °C for 12 h. TLC was used to monitor the progress of reaction. After full conversion, the mixture was cooled to room temperature and diluted with DCM (50 mL). Then it was washed with brine (2 × 50 mL), dried over MgSO$_4$ and the solvent was removed under reduced pressure. The crude product was purified by column chromatography on silica gel (cyclohexane/dichloromethane = 3:1 to 1:1) to yield the target compound 5,11-bis(3,5-dimethyl-[1,1'-biphenyl]-4-yl)-5,11-dihydroindolo[3,2-*b*]carbazole (115 mg, 186 µmol, 75%) as a colorless solid.

R$_f$ = 0.60 (cyclohexane/ethyl acetate = 10:1). – **^{1}H NMR** (400 MHz, CDCl$_3$) δ = 8.15 (d, *J* = 7.7 Hz, 2H), 7.77 (d, *J* = 7.6 Hz, 4H), 7.70 (s, 2H), 7.59 (s, 4H), 7.54 (t, *J* = 7.6 Hz, 4H), 7.48–7.33 (m, 4H), 7.22 (t, *J* = 7.5 Hz, 2H), 7.02 (d, *J* = 8.0 Hz, 2H), 2.03 (s, 12H) ppm. – **^{13}C NMR** (101 MHz, CDCl$_3$) δ = 141.6 (C$_q$, 4C$_{Ar}$), 140.8 (C$_q$, 2C$_{Ar}$), 139.0 (C$_q$, 4C$_{Ar}$), 136.5 (C$_q$, 2C$_{Ar}$), 134.5 (C$_q$, 2C$_{Ar}$), 129.0 (+, 4C$_{Ar}$H), 127.7 (+, 2C$_{Ar}$H), 127.6 (+, 4C$_{Ar}$H), 127.4 (+, 4C$_{Ar}$H), 126.2 (+, 2C$_{Ar}$H), 123.3 (C$_q$, 2C$_{Ar}$), 123.1 (C$_q$, 2C$_{Ar}$), 120.6 (+, 2C$_{Ar}$H), 118.9 (+, 2C$_{Ar}$H), 109.3 (+, 2C$_{Ar}$H), 99.9 (+, 2C$_{Ar}$H), 18.1 (+, 4CH$_3$) ppm. – **IR** (ATR, ṽ) = 2917, 1504, 1475, 1452, 1441, 1377, 1337, 1329, 1316, 1288, 1232, 1204, 1188, 1160, 1145, 1103, 1072, 1031, 1020, 1001, 986, 935, 926, 871, 856, 761, 745, 722, 694, 667, 646, 628, 612, 595, 567, 558, 547, 537, 487, 463, 416, 382 cm^{-1}. – **MS** (FAB, 3-NBA), *m/z* (%): 616 [M]$^+$. – **HRMS** (FAB, 3-NBA, C$_{46}$H$_{36}$N$_2$) calc.: 616.2878; found: 616.2877.

5,11-Bis(2,6-dimethyl-4-(5-methylpyridin-2-yl)phenyl)-5,11-dihydroindolo[3,2-b]carbazole (109)

Under argon atmosphere, a mixture of 5,11-bis(2,6-dimethyl-4-(4,4,5,5-tetramethyl-1,3,2-dioxaborolan-2-yl)phenyl)-5,11-dihydroindolo[3,2-b]carbazole (119 mg, 167 µmol, 1.00 equiv.), 2-bromo-5-methylpyridine (71.7 mg, 417 µmol, 2.50 equiv.), Pd(PPh$_3$)$_4$ (19.0 mg, 16.4 µmol, 10 mol%) and K$_2$CO$_3$ (230 mg, 1.66 mmol, 10.00 equiv.) in PhMe/EtOH/H$_2$O (8.00/1.00/1.00 mL) was stirred at 90 °C for 12 h. TLC was used to monitor the progress of reaction. After full conversion, the mixture was cooled to room temperature and diluted with DCM (50 mL). Then it was washed with brine (2 × 50 mL), dried over MgSO$_4$ and the solvent was removed under reduced pressure. The crude product was purified by column chromatography on silica gel (cyclohexane/dichloromethane = 2:1 to 1:1) to yield the target compound 5,11-bis(2,6-dimethyl-4-(5-methylpyridin-2-yl)phenyl)-5,11-dihydroindolo[3,2-b]carbazole (75.0 mg, 116 µmol, 70%) as a yellow luminescent solid

R_f = 0.25 (cyclohexane/ethyl acetate = 5:1). – **^{1}H NMR** (400 MHz, CDCl$_3$) δ = 8.62–8.61 (m, 2H), 8.11 (d, J = 7.7 Hz, 2H), 7.94 (s, 4H), 7.79 (d, J = 8.0 Hz, 2H), 7.65–7.67 (m, 4H), 7.40–7.36 (m, 2H), 7.22 (t, J = 7.5 Hz, 2H), 6.99 (d, J = 8.1 Hz, 2H), 2.45 (s, 6H), 2.04 (s, 12H) ppm. – **^{13}C NMR** (101 MHz, CDCl$_3$) δ = 154.6 (C$_q$, 2C$_{Ar}$), 150.3 (+, 2C$_{Ar}$H), 141.5 (C$_q$, 2C$_{Ar}$), 139.7 (C$_q$, 2C$_{Ar}$),139.1 (C$_q$, 4C$_{Ar}$), 137.7 (+, 2C$_{Ar}$H), 136.4 (C$_q$, 2C$_{Ar}$), 135.7 (C$_q$, 2C$_{Ar}$), 132.1 (C$_q$, 2C$_{Ar}$), 127.3 (+, 4C$_{Ar}$H), 126.2 (+, 2C$_{Ar}$H), 123.3 (C$_q$, 2C$_{Ar}$), 123.1 (C$_q$, 2C$_{Ar}$), 120.6 (+, 2C$_{Ar}$H), 120.5 (+, 2C$_{Ar}$H), 119.0 (+, 2C$_{Ar}$H), 109.3 (+, 2C$_{Ar}$H), 99.9 (+, 2C$_{Ar}$H), 18.4 (+, 2CH$_3$), 18.1 (+, 4CH$_3$) ppm. – **IR** (ATR, $\tilde{v}$) = 2921, 1612, 1596, 1562, 1506, 1472, 1453, 1443, 1374, 1339, 1329, 1317, 1290, 1231, 1204, 1190, 1164, 1142, 1105, 1084, 1028, 1003, 989, 938, 926, 841, 832, 796, 761, 747, 688, 649, 619, 548, 492, 414 cm^{-1}. – **MS** (FAB, 3-NBA), m/z (%): 647 [M]$^+$. – **HRMS** (FAB, 3-NBA, C$_{46}$H$_{39}$N$_4$) calc.: 647.3175 [M+H]$^+$; found: 647.3176 [M+H]$^+$.

5,11-Bis(2,6-dimethyl-4-(5-methylpyrimidin-2-yl)phenyl)-5,11-dihydroindolo[3,2-b]carbazole (110)

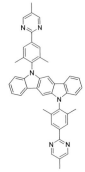

Under argon atmosphere, a mixture of 5,11-bis(2,6-dimethyl-4-(4,4,5,5-tetramethyl-1,3,2-dioxaborolan-2-yl)phenyl)-5,11-dihydroindolo[3,2-b]carbazole (119 mg, 166 μmol, 1.00 equiv.), 2-chloro-5-methylpyrimidine (53.0 mg, 412 μmol, 2.48 equiv.), Pd(PPh$_3$)$_4$ (19.0 mg, 16.4 μmol, 10 mol%) and K$_2$CO$_3$ (230 mg, 1.66 mmol, 10.0 equiv.) in PhMe/EtOH/H$_2$O (6.00/0.83/0.83 mL) was stirred at 90 °C for 12 h. TLC was used to monitor the progress of reaction. After full conversion, the mixture was cooled to room temperature and diluted with DCM (50 mL). Then it was washed with brine (2 × 50 mL), dried over MgSO$_4$ and the solvent was removed under reduced pressure. The crude product was purified by column chromatography on silica gel (cyclohexane/dichloromethane = 2:1 to 1:1) to yield the target compound 5,11-bis(2,6-dimethyl-4-(5-methylpyrimidin-2-yl)phenyl)-5,11-dihydroindolo[3,2-b]carbazole (86.0 mg, 133 μmol, 80%) as a yellow luminescent solid.

R$_f$ = 0.20 (cyclohexane/ethyl acetate = 5:1). – **^{1}H NMR** (400 MHz, CDCl$_3$) δ = 8.74 (s, 4H), 8.37 (s, 4H), 8.10 (d, J = 7.6 Hz, 2H), 7.66 (s, 2H), 7.38 (t, J = 8.1 Hz, 2H), 7.22 (t, J = 7.4 Hz, 2H), 6.98 (d, J = 8.1 Hz, 2H), 2.42 (s, 6H), 2.05 (s, 12H) ppm. – **^{13}C NMR** (101 MHz, CDCl$_3$) δ = 162.5 (C$_q$, 2C$_{Ar}$), 157.7 (+, 4C$_{Ar}$H), 141.4 (C$_q$, 2C$_{Ar}$), 139.1 (C$_q$, 4C$_{Ar}$), 137.9 (C$_q$, 2C$_{Ar}$), 137.4 (C$_q$, 2C$_{Ar}$), 136.3 (C$_q$, 2C$_{Ar}$), 128.8 (C$_q$, 2C$_{Ar}$), 128.4 (+, 4C$_{Ar}$H), 126.2 (+, 2C$_{Ar}$H), 123.3 (C$_q$, 2C$_{Ar}$), 123.2 (C$_q$, 2C$_{Ar}$), 120.5 (+, 2C$_{Ar}$H), 119.0 (+, 2C$_{Ar}$H), 109.3 (+, 2C$_{Ar}$H), 99.9 (+, 2C$_{Ar}$H), 18.1 (+, 4CH$_3$), 15.8 (+, 2CH$_3$) ppm. – **IR** (ATR, ṽ) = 2986, 1587, 1543, 1506, 1487, 1475, 1453, 1445, 1426, 1377, 1337, 1319, 1292, 1248, 1234, 1190, 1166, 1146, 1105, 1089, 1038, 1020, 1004, 994, 939, 928, 916, 894, 861, 851, 799, 762, 748, 720, 690, 653, 618, 595, 567, 558, 547, 492, 470, 443, 425, 416, 404, 394, 387 cm^{-1}. – **MS** (FAB, 3-NBA), m/z (%): 648 [M]$^+$. – **HRMS** (FAB, 3-NBA, C$_{44}$H$_{36}$N$_6$) calc.: 648.3001; found: 648.3004.

5,11-Bis(2,6-dimethyl-4-(5-(trifluoromethyl)pyrimidin-2-yl)phenyl)-5,11-dihydroindolo[3,2-b]carbazole (111)

Under argon atmosphere, a mixture of 5,11-bis(2,6-dimethyl-4-(4,4,5,5-tetramethyl-1,3,2-dioxaborolan-2-yl)phenyl)-5,11-dihydroindolo[3,2-*b*]carb-azole (119 mg, 166 µmol, 1.00 equiv.), 2-chloro-5-(trifluoromethyl)pyrimidine (76.0 mg, 416 µmol, 2.51 equiv.), Pd(PPh$_3$)$_4$ (19.0 mg, 16.4 µmol, 10 mol%) and K$_2$CO$_3$ (230 mg, 1.66 mmol, 10.0 equiv.) in PhMe/EtOH/H$_2$O (6.00/0.83/0.83 ml) was stirred at 90 °C for 12 h. TLC was used to monitor the progress of reaction. After full conversion, the mixture was cooled to room temperature and diluted with DCM (50 mL). Then it was washed with brine (2 × 50 mL), dried over MgSO$_4$ and the solvent was removed under reduced pressure. The crude product was purified by column chromatography on silica gel (cyclohexane/dichloromethane = 2:1 to 1:1) to yield the target compound 5,11-bis(2,6-dimethyl-4-(5-(trifluoromethyl)pyrimidin-2-yl)phenyl)-5,11-dihydroin-dolo[3,2-*b*]carb-azole (110 mg, 145 µmol, 88%) as a yellow luminescent solid.

R$_f$ = 0.20 (cyclohexane/ethyl acetate = 5:1). – **^{1}H NMR** (400 MHz, CDCl$_3$) δ = 9.13 (s, 4H), 8.50 (s, 4H), 8.11 (d, *J* = 7.7 Hz, 2H), 7.66 (s, 2H), 7.39 (t, *J* = 7.6 Hz, 2H), 7.23 (t, *J* = 7.4 Hz, 2H), 6.98 (d, *J* = 8.1 Hz, 2H), 2.08 (s, 12H) ppm. – **^{13}C NMR** (101 MHz, CDCl$_3$) δ = 167.3 (C$_q$, 2C$_{Ar}$), 154.8 (+, q, *J* = 3.2 Hz, 4C$_{Ar}$H), 141.3 (C$_q$, 2C$_{Ar}$), 139.5 (C$_q$, 4C$_{Ar}$), 138.9 (C$_q$, 2C$_{Ar}$), 136.5 (C$_q$, 2C$_{Ar}$), 136.2 (C$_q$, 2C$_{Ar}$), 129.5 (+, 4C$_{Ar}$H), 126.4 (+, 2C$_{Ar}$H), 123.4 (C$_q$, 2C$_{Ar}$), 123.3 (q, *J* = 272.0 Hz, 2CF$_3$), 123.2 (C$_q$, 2C$_{Ar}$), 122.8 (C$_q$, q, *J* = 33.9 Hz, 2C$_{Ar}$), 120.6 (+, 2C$_{Ar}$H), 119.2 (+, 2C$_{Ar}$H), 109.3 (+, 2C$_{Ar}$H), 99.9 (+, 2C$_{Ar}$H), 18.1 (+, 4CH$_3$) ppm. – **^{19}F NMR** (376 MHz, CDCl$_3$) δ = –66.50 ppm. – **IR** (ATR, ṽ) = 2929, 1594, 1547, 1506, 1476, 1442, 1381, 1373, 1322, 1292, 1235, 1228, 1187, 1167, 1140, 1125, 1108, 1081, 1017, 952, 939, 929, 807, 761, 741, 418 cm^{-1}. – **MS** (FAB, 3-NBA), *m/z* (%): 756 [M]$^+$. – **HRMS** (FAB, 3-NBA, C$_{44}$H$_{30}$N$_6$F$_6$) calc.: 756.2436; found: 756.2434.

5,11-Bis(2,6-dimethyl-4-(5-(trifluoromethyl)pyrazin-2-yl)phenyl)-5,11-dihydroindolo[3,2-b]carbazole (112)

Under argon atmosphere, a mixture of 5,11-bis(2,6-dimethyl-4-(4,4,5,5-tetramethyl-1,3,2-dioxaborolan-2-yl)phenyl)-5,11-dihydroindolo[3,2-*b*]carbazole (119 mg, 166 μmol, 1.00 equiv.), 2-chloro-5-(trifluoromethyl)pyrazine (76.0 mg, 416 μmol, 2.51 equiv.), Pd(PPh$_3$)$_4$ (19.0 mg, 16.4 μmol, 10 mol%) and K$_2$CO$_3$ (230 mg, 1.66 mmol, 10.0 equiv.) in PhMe/EtOH/H$_2$O (6.00/0.83/0.83 ml) was stirred at 90 °C for 12 h. TLC was used to monitor the progress of reaction. After full conversion, the mixture was cooled to room temperature and diluted with DCM (50 mL). Then it was washed with brine (2 × 50 mL), dried over MgSO$_4$ and the solvent was removed under reduced pressure. The crude product was purified by column chromatography on silica gel (cyclohexane/dichloromethane = 2:1 to 1:1) to yield the target compound 5,11-bis(2,6-dimethyl-4-(5-(trifluoromethyl)pyrazin-2-yl)phenyl)-5,11-dihydroindolo[3,2-*b*]carbazole (98.0 mg, 130 μmol, 78%) as a yellow luminescent solid.

R_f = 0.20 (cyclohexane/ethyl acetate = 5:1). – **^{1}H NMR** (400 MHz, CDCl$_3$) δ = 9.27 (s, 2H), 9.09 (s, 2H), 8.12 (d, J = 7.8 Hz, 2H), 8.07 (s, 4H), 7.65 (s, 2H), 7.40 (t, J = 7.6 Hz, 2H), 7.26–7.22 (t, J = 8.0 Hz, 2H), 6.98 (d, J = 8.1 Hz, 2H), 2.09 (s, 12H) ppm. – **^{13}C NMR** (101 MHz, CDCl$_3$) δ = 155.4 (C$_q$, 2C$_{Ar}$), 142.3 (C$_q$, 2C$_{Ar}$), 141.9 (+, 2C$_{Ar}$H), 141.3 (+, q, J = 3.1 Hz, 2C$_{Ar}$H), 140.1 (C$_q$, 4C$_{Ar}$), 138.1 (C$_q$, 2C$_{Ar}$), 136.2 (C$_q$, 2C$_{Ar}$), 135.4 (C$_q$, 2C$_{Ar}$), 128.7 (C$_q$, q, J = 81.6 Hz, 2C$_{Ar}$), 128.1 (+, 4C$_{Ar}$H), 126.5 (+, 2C$_{Ar}$H), 123.5 (C$_q$, 2C$_{Ar}$), 123.2 (C$_q$, 2C$_{Ar}$), 121.6 (q, J = 274.0 Hz, 2CF$_3$), 120.6 (+, 2C$_{Ar}$H), 119.3 (+, 2C$_{Ar}$H), 109.2 (+, 2C$_{Ar}$H), 99.9 (+, 2C$_{Ar}$H), 18.2 (+, 4CH$_3$) ppm. – **^{19}F NMR** (376 MHz, CDCl$_3$) δ = –71.63 ppm. – **IR** (ATR, ṽ) = 2925, 1575, 1528, 1507, 1476, 1453, 1443, 1371, 1341, 1332, 1320, 1299, 1292, 1272, 1249, 1235, 1220, 1187, 1154, 1132, 1101, 1079, 1037, 1021, 1003, 993, 958, 938, 926, 890, 874, 853, 847, 837, 788, 759, 739, 711, 688, 659, 632, 596, 548, 482, 470, 416, 401 cm^{-1}. – **MS** (FAB, 3-NBA), m/z (%): 756 [M]$^+$. – **HRMS** (FAB, 3-NBA, C$_{44}$H$_{30}$N$_6$F$_6$) calc.: 756.2436; found: 756.2435.

5,11-Bis(4-(4,6-bis(4-(*tert*-butyl)phenyl)-1,3,5-triazin-2-yl)-2,6-dimethylphenyl)-5,11-dihydroindolo[3,2-b]carbazole (113)

Under argon atmosphere, a mixture of 5,11-bis(2,6-dimethyl-4-(4,4,5,5-tetramethyl-1,3,2-dioxaborolan-2-yl)phenyl)-5,11-dihydroindolo[3,2-*b*]carbazole (119 mg, 166 μmol, 1.00 equiv.), 2,4-bis(4-tert-butylphenyl)-6-chloro-1,3,5-triazine (158 mg, 416 μmol, 2.50 equiv.), Pd(PPh₃)₄ (19.0 mg, 16.4 μmol, 10 mol%) and K₂CO₃ (230 mg, 1.66 mmol, 10.0 equiv.) in PhMe/EtOH/H₂O (8.00/1.00/1.00 mL) was stirred at 90 °C for 12 h. TLC was used to monitor the progress of reaction. After full conversion, the mixture was cooled to room temperature and diluted with DCM (50 mL). Then it was washed with brine (2 × 50 mL), dried over MgSO₄ and the solvent was removed under reduced pressure. The crude product was purified by column chromatography on silica gel (cyclohexane/dichloromethane = 5:1 to 2:1) to yield the target compound 5,11-bis(4-(4,6-bis(4-(tert-butyl)phenyl)-1,3,5-triazin-2-yl)-2,6-dimethylphenyl)-5,11-dihydroindolo[3,2-*b*]carbazole (125 mg, 109 μmol, 65%) as a yellow luminescent solid.

R_f = 0.50 (cyclohexane/ethyl acetate = 20:1). – **¹H NMR** (400 MHz, CDCl₃) δ = 8.84–8.67 (m, 12H), 8.14 (d, J = 7.8 Hz, 2H), 7.70 (s, 2H), 7.65 (d, J = 8.6 Hz, 8H), 7.41 (t, J = 7.6 Hz, 2H), 7.25 (d, J = 8.0 Hz, 2H), 7.02 (d, J = 8.1 Hz, 2H), 2.14 (s, 12H), 1.44 (s, 36H) ppm. – **¹³C NMR** (101 MHz, CDCl₃) δ = 171.9 (C$_q$, 4C$_{Ar}$), 171.4 (C$_q$, 2C$_{Ar}$), 156.4 (C$_q$, 4C$_{Ar}$), 141.3 (C$_q$, 2C$_{Ar}$), 139.2 (C$_q$, 4C$_{Ar}$), 139.1 (C$_q$, 2C$_{Ar}$), 136.8 (C$_q$, 2C$_{Ar}$), 136.3 (C$_q$, 4C$_{Ar}$), 133.7 (C$_q$, 2C$_{Ar}$), 129.5 (+, 4C$_{Ar}$H), 129.0 (+, 8C$_{Ar}$H), 126.4 (+, 2C$_{Ar}$H), 125.9 (+, 8C$_{Ar}$H), 123.4 (C$_q$, 2C$_{Ar}$), 123.2 (C$_q$, 2C$_{Ar}$), 120.6 (+, 2C$_{Ar}$H), 119.2 (+, 2C$_{Ar}$H), 109.3 (+, 2C$_{Ar}$H), 100.0 (+, 2C$_{Ar}$H), 35.3 (C$_q$, 4C(CH₃)₃), 31.4 (+, 12CH₃) , 18.2 (+, 4CH₃) ppm. – **IR** (ATR, $\tilde{v}$) = 2962 (w), 1578 (w), 1517 (vs), 1487 (s), 1476 (s), 1453 (s), 1445 (s), 1419 (m), 1409 (s), 1366 (vs), 1337 (m), 1322 (m), 1307 (w), 1290 (w), 1265 (w), 1254 (m), 1234 (s), 1191 (m), 1147 (w), 1105 (m), 1017 (w), 857 (w), 816 (vs), 762 (w), 738 (s), 713 (w), 704 (w), 578 (m), 567 (w), 551 (w) cm⁻¹. – **HRMS** (ESI, C₈₀H₇₈N₈) calc.: 1150.6349, found 1150.6324.

5.4 Analytical Data of [2,2]Paracyclophane-Based Donors for TADF Emitters

(*rac*)-4-*N*-(2-Chlorophenyl)amino[2.2]paracyclophane (140)

Under argon atmosphere, a mixture of (*rac*)-4-bromo[2.2]paracyclophane (574 mg, 2.00 mmol, 1.00 equiv.), 2-chloroaniline (281 mg, 2.20 mmol, 1.10 equiv.), Pd(OAc)$_2$ (22.4 mg, 99.8 µmol, 5 mol%), tri-*tert*-butylphosphonium tetrafluoroborate (29.0 mg, 100.0 µmol, 5 mol%) and sodium *tert*-butoxide (240 mg, 2.50 mmol, 1.25 equiv.) in toluene (10.0 mL) was stirred at 100 °C for 12 h. The mixture was cooled to room temperature and diluted with dichloromethane (20 mL). Then it was washed with brine (3 × 20 mL). The organic layer was dried over MgSO$_4$ and the solvent was removed under reduced pressure. The obtained crude product was purified *via* column chromatography on silica gel (cyclohexane/dichloromethane = 5:1 to 3:1) to yield the title compound as a yellow solid (510 mg, 1.53 mmol, 76%).

R$_f$ = 0.60 (cyclohexane/dichloromethane = 3:1). – **^{1}H NMR** (400 MHz, CDCl$_3$) δ = 7.39 (d, *J* = 8.6 Hz, 1H), 7.13 (d, *J* = 7.8 Hz, 1H), 7.05 (t, *J* = 8.1 Hz, 1H), 6.91 (d, *J* = 8.1 Hz, 1H), 6.77 (t, *J* = 8.0 Hz, 1H), 6.59 (d, *J* = 7.7 Hz, 1H), 6.53–6.43 (m, 4H), 5.98 (s, 1H), 5.95 (s, 1H), 3.13–2.89 (m, 7H), 2.76–2.64 (m, 1H) ppm. – **^{13}C NMR** (101 MHz, CDCl$_3$) δ = 141.6 (C$_q$, C$_{Ar}$), 140.2 (C$_q$, C$_{Ar}$), 139.6 (C$_q$, C$_{Ar}$), 139.1 (C$_q$, C$_{Ar}$), 138.9 (C$_q$, C$_{Ar}$), 136.1 (+, C$_{Ar}$H), 133.8 (+, C$_{Ar}$H), 132.9 (+, C$_{Ar}$H), 132.9 (C$_q$, C$_{Ar}$), 131.3 (+, C$_{Ar}$H), 129.5 (+, C$_{Ar}$H), 128.1 (+, C$_{Ar}$H), 127.9 (+, C$_{Ar}$H), 127.5 (+, C$_{Ar}$H), 127.4 (+, C$_{Ar}$H), 120.4 (+, C$_{Ar}$H), 119.7 (+, C$_{Ar}$H), 114.4 (+, C$_{Ar}$H), 35.3 (–, CH$_2$), 34.9 (–, CH$_2$), 34.0 (–, CH$_2$), 33.8 (–, CH$_2$) ppm. – **IR** (ATR, ṽ) = 3391 (w), 2922 (w), 2888 (w), 2850 (w), 1591 (vs), 1562 (w), 1509 (s), 1489 (vs), 1458 (m), 1435 (vs), 1412 (w), 1312 (vs), 1289 (w), 1261 (w), 1228 (w), 1181 (w), 1159 (w), 1149 (w), 1126 (w), 1091 (w), 1050 (w), 1033 (s), 983 (w), 941 (w), 898 (m), 885 (w), 846 (w), 817 (w), 799 (m), 739 (vs), 718 (s), 688 (m), 656 (m), 642 (w), 630 (w), 591 (w), 579 (w), 514 (m), 503 (m), 455 (m), 436 (s), 409 (m), 401 (m), 391 (m), 378 (m) cm^{-1}. – **MS** (EI, 70 °C), *m/z* (%): 333 [M]$^+$. – **HRMS** (EI) calc. for C$_{22}$H$_{20}$N$_1$Cl$_1$ [M]$^+$ 333.1284, found 333.1286.

(*rac*)-[2]Paracyclo[2](1,4)carbazolophane (130, Carbazolophane, Czp)

Under argon atmosphere, a mixture of (*rac*)-4-*N*-(2-chlorophenyl)amino[2.2]paracyclophane (6.00 g, 18.0 mmol, 1.00 equiv.), Pd$_2$(dba)$_3$ (1.64 g, 1.79 mmol, 10 mol%), 2-dicyclohexylphosphino-2',4',6'-triisopropylbiphenyl (XPhos, 2.57 g, 5.39 mmol, 30 mol%), pivalic acid (1.10 g, 10.8 mmol, 50 mol%) and potassium carbonate (12.4 g, 89.9 mmol, 5.00 equiv.) in anhydrous *N*,*N*-dimethylacetamide (50 mL) was stirred at 110 °C for 12 h. The mixture was cooled to room temperature and diluted with dichloromethane (100 mL). Then it was washed with brine (3 × 100 mL). The organic layer was dried over MgSO$_4$ and the solvent was removed under reduced pressure. The obtained crude product was purified *via* column chromatography on silica gel (cyclohexane/dichloromethane = 5:1 to 2:1) to yield the title compound as a white solid (3.25 g, 10.9 mmol, 61%).

R$_f$ = 0.15 (cyclohexane/dichloromethane = 2:1). – **^{1}H NMR** (400 MHz, CDCl$_3$) δ = 8.06 (d, *J* = 7.9 Hz, 1H), 7.87 (bs, NH, 1H), 7.47 (d, *J* = 8.0 Hz, 1H), 7.40 (t, *J* = 8.0 Hz, 1H), 7.31–7.20 (m, 1H), 6.63 (d, *J* = 7.5 Hz, 1H), 6.56 (d, *J* = 7.5 Hz, 1H), 6.50 (dd, *J* = 7.8 Hz, *J* = 1.7 Hz, 1H), 6.36 (dd, *J* = 7.8 Hz, *J* = 1.9 Hz, 1H), 5.94 (dd, *J* = 7.7 Hz, *J* = 1.9 Hz, 1H), 5.23 (dd, *J* = 7.8 Hz, *J* = 2.0 Hz, 1H), 4.04–3.98 (m, 1H), 3.40–3.30 (m, 1H), 3.16–2.91 (m, 6H) ppm. – **^{13}C NMR** (101 MHz, CDCl$_3$) δ = 140.1 (C$_q$, C$_{Ar}$), 138.9 (C$_q$, C$_{Ar}$), 138.0 (C$_q$, C$_{Ar}$), 137.4 (C$_q$, C$_{Ar}$), 135.9 (C$_q$, C$_{Ar}$), 132.1 (+, C$_{Ar}$H), 131.6 (+, C$_{Ar}$H), 131.1 (+, C$_{Ar}$H), 126.6 (+, C$_{Ar}$H), 126.4 (+, C$_{Ar}$H), 125.5 (C$_q$, C$_{Ar}$), 125.3 (C$_q$, C$_{Ar}$), 125.0 (+, C$_{Ar}$H), 124.7 (+, C$_{Ar}$H), 122.5 (+, C$_{Ar}$H), 122.3 (C$_q$, C$_{Ar}$), 119.8 (+, C$_{Ar}$H), 110.8 (+, C$_{Ar}$H), 34.0 (–, CH$_2$), 33.9 (–, CH$_2$), 33.3 (–, CH$_2$), 31.2 (–, CH$_2$) ppm. – **IR** (ATR, ṽ) = 3391 (m), 2921 (w), 2850 (vw), 1606 (vw), 1592 (w), 1565 (vw), 1509 (vw), 1496 (w), 1458 (w), 1438 (w), 1407 (vw), 1394 (w), 1323 (w), 1302 (w), 1248 (w), 1027 (w), 935 (w), 868 (w), 802 (w), 751 (vs), 737 (s), 720 (m), 674 (w), 611 (w), 582 (w), 516 (m), 470 (m), 441 (w), 401 (w) cm^{-1}. – **MS** (EI, 130 °C), *m/z* (%): 297 [M]$^+$. – **HRMS** (EI, 130 °C) calc. for C$_{22}$H$_{19}$N$_1$ [M]$^+$ 297.1517, found 297.1519. The analytical data is consistent with literature.[78]

(*rac*)-[2]Paracyclo[2]6-(bromo)(1,4)carbazolophane (141)

A solution of NBS (201 mg, 1.10 mmol, 1.10 equiv.) in 3.0 mL of THF was added slowly to a solution of (*rac*)-[2]paracyclo[2](1,4)carbazolophane (305 mg, 1.00 mmol, 1.00 equiv.) in 6 mL of THF in a round bottom flask at 0 °C. The reaction mixture was stirred at this temperature for 1 h. The mixture was quenched with water and diluted with dichloromethane (20 mL). Then it was washed with brine (3 × 20 mL). The organic layer was dried over $MgSO_4$ and the solvent was removed under reduced pressure. The obtained crude product was purified *via* column chromatography on silica gel (cyclohexane/dichloromethane = 5:1 to 2:1) to yield the title compound as a white solid (321 mg, 0.853 mmol, 83%).

R_f = 0.50 (cyclohexane/dichloromethane = 1:1). – **^{1}H NMR** (400 MHz, CDCl$_3$) δ = 8.17 (d, *J* = 1.7 Hz, 1H), 7.87 (bs, NH, 1H), 7.49 (dd, *J* = 8.6 Hz, *J* = 1.9 Hz, 1H), 7.34 (d, *J* = 8.5 Hz, 1H), 6.65 (d, *J* = 7.5 Hz, 1H), 6.58 (d, *J* = 7.5 Hz, 1H), 6.51 (dd, *J* = 7.8 Hz, *J* = 1.7 Hz, 1H), 6.37 (dd, *J* = 7.8 Hz, *J* = 1.9 Hz, 1H), 5.93 (dd, *J* = 7.8 Hz, *J* = 1.9 Hz, 1H), 5.30 (dd, *J* = 7.8 Hz, *J* = 1.8 Hz, 1H), 3.97–3.88 (m, 1H), 3.40–3.24 (m, 1H), 3.23–2.88 (m, 6H) ppm. – **^{13}C NMR** (101 MHz, CDCl$_3$) δ = 140.6 (C_q, C_{Ar}), 138.0 (C_q, C_{Ar}), 137.5 (C_q, C_{Ar}), 137.4 (C_q, C_{Ar}), 136.1 (C_q, C_{Ar}), 132.3 (C_q, C_{Ar}), 131.8 (+, C_{Ar}H), 131.8 (+, C_{Ar}H), 127.7 (+, C_{Ar}H), 127.1 (C_q, C_{Ar}), 127.0 (+, C_{Ar}H), 126.5 (+, C_{Ar}H), 124.9 (+, C_{Ar}H), 124.7 (+, C_{Ar}H), 124.6 (C_q, C_{Ar}), 122.5 (+, C_{Ar}H), 112.5 (C_q, C_{Ar}), 112.2 (+, C_{Ar}H), 34.0 (–, CH_2), 33.7 (–, CH_2), 33.3 (–, CH_2), 31.2 (–, CH_2) ppm. – **MS** (FAB, 3-NBA), *m/z* (%): 375 [M]$^+$. – **HRMS** (FAB, 3-NBA) calc. for $C_{22}H_{18}N_1{}^{79}Br_1$ [M]$^+$ 375.0623, found 375.0621. The analytical data is consistent with literature.[18]

(*rac*)-[2]Paracyclo[2]6-(cyano)(1,4)carbazolophane (142)

A mixture of (*rac*)-[2]paracyclo[2]6-(bromo)(1,4)carbazolophane (105 mg, 0.279 mmol, 1.00 equiv.) and copper(I) cyanid (53.0 mg, 0.592 mmol, 2.12 equiv.) in DMF (5 mL) was heated to 150 °C and stirred for 48 h. After cooling to room temperature, the reaction mixture was diluted with 50 mL of dichloromethane and washed with saturated aqueous Na_2CO_3 (3 × 100 mL). The organic layer was dried over $MgSO_4$ and the solvent was removed under reduced pressure. The obtained crude product was purified *via* column chromatography on silica gel (cyclohexane/dichloromethane = 5:1 to 1:1) to yield the title compound as a white solid (80.0 mg, 0.248 mmol, 89%).

R_f = 0.20 (cyclohexane/dichloromethane = 1:1). – **^{1}H NMR** (400 MHz, CDCl$_3$) δ = 8.37 (s, 1H), 8.21 (bs, NH, 1H), 7.67 (dd, *J* = 8.4 Hz, *J* = 1.5 Hz, 1H), 7.53 (dd, *J* = 8.4 Hz, *J* = 0.6 Hz, 1H), 6.72 (d, *J* = 7.6 Hz, 1H), 6.65 (d, *J* = 7.5 Hz, 1H), 6.52 (dd, *J* = 7.9, *J* = 1.9 Hz, 1H), 6.38 (dd, *J* = 7.8 Hz, *J* = 2.0 Hz, 1H), 5.91 (dd, *J* = 7.8 Hz, *J* = 2.0 Hz, 1H), 5.22 (dd, *J* = 7.8 Hz, *J* = 1.9 Hz, 1H), 4.05–3.84 (m, 1H), 3.44–3.26 (m, 1H), 3.25–2.78 (m, 6H) ppm. – **^{13}C NMR** (101 MHz, CDCl$_3$) δ = 140.8 (C_q, C_{Ar}), 138.1 (C_q, C_{Ar}), 137.6 (C_q, C_{Ar}), 136.4 (C_q, C_{Ar}), 132.7 (+, C_{Ar}H), 132.6 (+, C_{Ar}H), 132.2 (+, C_{Ar}H), 128.5 (+, C_{Ar}H), 128.1 (+, C_{Ar}H), 127.5 (+, C_{Ar}H), 126.5 (+, C_{Ar}H), 125.4 (C_q, C_{Ar}), 124.9 (+, C_{Ar}H), 124.6 (C_q, C_{Ar}), 122.9 (C_q, C_{Ar}), 121.0 (C_q, C_{Ar}), 111.6 (+, C_{Ar}H), 102.7 (C_q, C_{Ar}), 34.1 (–, CH_2), 33.8 (–, CH_2), 33.3 (–, CH_2), 31.2 (–, CH_2) ppm. – **IR** (ATR, ṽ) = 3301 (m), 2934 (m), 2919 (m), 2220 (vs), 1595 (m), 1572 (m), 1468 (m), 1407 (w), 1307 (vs), 1261 (m), 1244 (w), 1215 (w), 1174 (m), 1130 (m), 892 (w), 875 (m), 809 (vs), 795 (s), 773 (m), 768 (m), 739 (w), 717 (m), 637 (s), 629 (m), 618 (vs), 569 (m), 520 (s), 514 (s), 499 (s) cm^{-1}. – **MS** (EI, 160°C), *m/z* (%): 322 [M]$^+$. – **HRMS** (EI, 160°C) calc. for $C_{23}H_{18}N_2$ [M]$^+$ 322.1470, found 322.1471.

(*rac*)-1-(*N*-[2]Paracyclo[2]-6-(cyano)(1,4)carbazolophanyl)-4-(4,6-diphenyl-1,3,5-triazin-2-yl)-benzene (132)

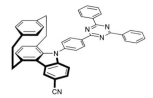

A 20 mL sealable vial was charged with (*rac*)-[2]paracyclo[2]6-(cyano)(1,4)carbazolophane (120 mg, 372 μmol, 1.00 equiv.), 2-(4-fluorophenyl)-4,6-diphenyl-1,3,5-triazine (146 mg, 447 μmol, 1.20 equiv.) and potassium phosphate tribasic (395 mg, 1.86 mmol, 5.00 equiv.). It was evacuated and flushed with argon three times. Through the septum 8 mL of anhydrous DMSO were added, then it was heated to 150 °C and stirred for 12 h. After cooling to room temperature, the reaction mixture was diluted with 50 mL of dichloromethane and washed with brine (3 × 50 mL). The organic layer was dried over MgSO$_4$ and the solvent was removed under reduced pressure. The obtained crude product was purified *via* column chromatography on silica gel (cyclohexane/dichloromethane = 2:1 to 1:1) to yield the title compound as a white solid (91.0 mg, 145 μmol, 39%).

R_f = 0.60 (cyclohexane/dichloromethane = 1:2). – **^{1}H NMR** (400 MHz, CDCl$_3$) δ = 9.05 (bs, 2H), 8.90–8.75 (m, 4H), 8.47 (d, *J* = 1.5 Hz, 1H), 8.08 (s, 1H), 7.75–7.37 (m, 9H), 6.82–6.70 (m, 2H), 6.53 (dd, *J* = 7.9, 1.9 Hz, 1H), 6.36 (dd, *J* = 7.9 Hz, *J* = 1.8 Hz, 1H), 5.95 (dd, *J* = 7.8 Hz, *J* = 1.9 Hz, 1H), 5.52 (dd, *J* = 7.8 Hz, *J* = 1.9 Hz, 1H), 4.1–3.92 (m, 1H), 3.32–3.15 (m, 2H), 3.13–3.03 (m, 1H), 2.99–2.83 (m, 1H), 2.78–2.70 (m, 2H), 2.36–2.23 (m, 1H) ppm. – **^{13}C NMR** (101 MHz, CDCl$_3$) δ = 172.0 (C$_q$, C$_{Ar}$), 170.7 (C$_q$, C$_{Ar}$), 142.2 (C$_q$, C$_{Ar}$), 141.6 (C$_q$, C$_{Ar}$), 137.5 (C$_q$, C$_{Ar}$), 136.3 (C$_q$, C$_{Ar}$), 136.1 (C$_q$, C$_{Ar}$), 136.1 (C$_q$, C$_{Ar}$), 135.1 (+, C$_{Ar}$H), 132.9 (+, C$_{Ar}$H), 132.2 (+, C$_{Ar}$H), 131.8 (+, C$_{Ar}$H), 130.6 (+, C$_{Ar}$H), 129.2 (+, C$_{Ar}$H), 128.9 (+, C$_{Ar}$H), 128.6 (+, C$_{Ar}$H), 127.4 (+, C$_{Ar}$H), 126.6 (+, C$_{Ar}$H), 125.8 (C$_q$, C$_{Ar}$), 125.8 (C$_q$, C$_{Ar}$), 125.7 (+, C$_{Ar}$H), 124.9 (C$_q$, C$_{Ar}$), 120.7 (C$_q$, C$_{Ar}$), 110.8 (+, C$_{Ar}$H), 103.7 (C$_q$, C$_{Ar}$), 35.1 (–, CH$_2$), 33.6 (–, CH$_2$), 33.4 (–, CH$_2$), 33.2 (–, CH$_2$) ppm. – **IR** (ATR, ṽ) = 2925 (w), 2854 (w), 2220 (w), 1596 (w), 1588 (w), 1509 (vs), 1462 (s), 1442 (s), 1411 (m), 1392 (m), 1361 (vs), 1299 (s), 1262 (s), 1244 (m), 1174 (m), 1146 (m), 1014 (m), 837 (m), 833 (m), 802 (s), 772 (s), 762 (vs), 735 (vs), 687 (vs), 667 (s), 660 (s), 645 (s), 640 (s), 615 (s), 592 (s), 584 (m), 562 (m), 517 (vs), 496 (s), 487 (m), 467 (m), 456 (m), 401 (m) cm^{-1}. – **MS** (FAB, 3-NBA), *m/z* (%): 630 [M+H]$^+$, 629 [M]$^+$. – **HRMS** (FAB, 3-NBA) calc. for C$_{44}$H$_{32}$N$_5$ [M+H]$^+$ 630.2658, found 630.2660.

(*rac*)-4-*N*-(2-Chloro-4-(trifluoromethyl)phenyl)amino[2.2]paracyclophane (144)

Under argon atmosphere, a mixture of (*rac*)-4-bromo[2.2]paracyclophane (287 mg, 1.00 mmol, 1.00 equiv.), 2-chloro-4-(trifluoromethyl)aniline (235 mg, 1.20 mmol, 1.20 equiv.), Pd$_2$(dba)$_3$ (45.8 mg, 50.0 μmol, 5 mol%), 2-dicyclohexylphosphino-2',4',6'-triisopropylbiphenyl (XPhos, 47.6 mg, 99.8 μmol, 10 mol%) and sodium *tert*-butoxide (144 mg, 1.50 mmol, 1.50 equiv.) in toluene (5 mL) was stirred at 100 °C for 12 h. The mixture was cooled to room temperature and diluted with dichloromethane (20 mL). Then it was washed with brine (3 × 20 mL). The organic layer was dried over MgSO$_4$ and the solvent was removed under reduced pressure. The obtained crude product was purified *via* column chromatography on silica gel (cyclohexane/dichloromethane = 10:1 to 7.5:1) to yield the title compound as a white solid (220 mg, 547 μmol, 55%).

R_f = 0.50 (cyclohexane/dichloromethane = 5:1). – **^{1}H NMR** (400 MHz, CDCl$_3$) δ = 7.64 (d, *J* = 2.1 Hz, 1H), 7.33–7.22 (m, 1H), 7.05 (dd, *J* = 7.9 Hz, *J* = 2.0 Hz, 1H), 6.87 (d, *J* = 8.6 Hz, 1H), 6.62–6.45 (m, 5H), 6.19 (s, 1H), 5.98 (s, 1H), 3.15–2.88 (m, 7H), 2.78–2.66 (m, 1H) ppm. – **^{13}C NMR** (101 MHz, CDCl$_3$) δ = 143.3 (C$_q$, C$_{Ar}$), 141.9 (C$_q$, C$_{Ar}$), 139.7 (C$_q$, C$_{Ar}$), 139.5 (C$_q$, C$_{Ar}$), 139.2 (C$_q$, C$_{Ar}$), 137.5 (C$_q$, C$_{Ar}$), 136.2, 134.2 (C$_q$, C$_{Ar}$), 133.9 (+, C$_{Ar}$H), 133.2 (+, C$_{Ar}$H), 133.1 (+, C$_{Ar}$H), 131.4 (+, C$_{Ar}$H), 129.6 (+, C$_{Ar}$H), 129.3 (+, C$_{Ar}$H), 127.2 (+, C$_{Ar}$H), 126.8 (+, q, *J* = 3.9 Hz, C$_{Ar}$H), 124.9 (+, q, *J* = 3.7 Hz, C$_{Ar}$H), 124.1 (C$_q$, q, *J* = 270.9 Hz, CF$_3$), 121.0 (C$_q$, q, *J* = 33.4 Hz, C$_{Ar}$), 119.4 (C$_q$, C$_{Ar}$), 113.1 (+, C$_{Ar}$H), 35.8 (–, CH$_2$), 35.3 (–, CH$_2$), 34.9 (–, CH$_2$), 33.9 (–, CH$_2$) ppm. – **^{19}F NMR** (101 MHz, CDCl$_3$) δ = –65.71 ppm. – **MS** (FAB, 3-NBA), *m/z* (%): 401 [M]$^+$. – **HRMS** (FAB, 3-NBA) calc. for C$_{23}$H$_{19}$^{35}Cl$_1$F$_3$ [M]$^+$ 401.1158, found 401.1158.

(*rac*)-[2]Paracyclo[2]6-(trifluoromethyl)(1,4)carbazolophane (143)

Under argon atmosphere, a mixture of (*rac*)-4-*N*-(2-chloro-4-(trifluoromethyl)phenyl)amino[2.2]paracyclophane (1.10 g, 2.70 mmol, 1.00 equiv.), Pd$_2$(dba)$_3$ (0.248 g, 0.271 mmol, 10 mol%), 2-dicyclohexylphosphino-2',4',6'-triisopropylbiphenyl (XPhos, 0.388 g, 0.813 mmol, 30 mol%), pivalic acid (0.166 g, 1.60 mmol, 60 mol%) and potassium carbonate (1.87 g, 14.0 mmol, 5.00 equiv.) in anhydrous *N,N*-dimethylacetamide (25 mL) was stirred at 110 °C for 12 h. Then the mixture was cooled to room temperature and diluted with dichloromethane (50 mL). Then it was washed with brine (3 × 50 mL). The organic layer was dried over MgSO$_4$ and the solvent was removed under reduced pressure. The obtained crude product was purified *via* column chromatography on silica gel (cyclohexane/dichloromethane = 5:1 to 3:1) to yield the title compound as a white solid (0.618 g, 1.70 mmol, 62%).

R_f = 0.20 (cyclohexane/dichloromethane = 2:1). – ^{1}**H NMR** (400 MHz, CDCl$_3$) δ = 8.31 (s, 1H), 8.14 (bs, NH, 1H), 7.65 (dd, *J* = 8.5 Hz, *J* = 1.7 Hz, 1H), 7.54 (d, *J* = 8.5 Hz, 1H), 6.70 (d, *J* = 7.5 Hz, 1H), 6.63 (d, *J* = 7.5 Hz, 1H), 6.52 (dd, *J* = 7.8 Hz, *J* = 1.9 Hz, 1H), 6.38 (dd, *J* = 7.8 Hz, *J* = 2.0 Hz, 1H), 5.92 (dd, *J* = 7.8 Hz, *J* = 2.0 Hz, 1H), 5.21 (dd, *J* = 7.8 Hz, *J* = 2.0 Hz, 1H), 4.00 (dd, *J* = 12.5 Hz, *J* = 9.8 Hz, 1H), 3.43–3.29 (m, 1H), 3.23–2.98 (m, 5H), 2.97–2.88 (m, 1H) ppm. – ^{13}**C NMR** (101 MHz, CDCl$_3$) δ = 140.8 (C$_q$, C$_{Ar}$), 140.4 (C$_q$, C$_{Ar}$), 138.0 (C$_q$, C$_{Ar}$), 137.5 (C$_q$, C$_{Ar}$), 136.2 (C$_q$, C$_{Ar}$), 132.4 (+, C$_{Ar}$H), 132.1 (+, C$_{Ar}$H), 131.9 (+, C$_{Ar}$H), 127.4 (+, C$_{Ar}$H), 126.4 (+, C$_{Ar}$H), 125.5 (C$_q$, q, *J* = 271.3 Hz, CF$_3$), 125.1 (C$_q$, C$_{Ar}$), 124.8 (C$_q$, C$_{Ar}$), 124.7 (+, C$_{Ar}$H), 122.7 (C$_q$, C$_{Ar}$), 122.02 (C$_q$, q, *J* = 31.9 Hz, C$_{Ar}$), 121.95 (+, q, *J* = 3.6 Hz, C$_{Ar}$H), 119.81 (+, q, *J* = 4.1 Hz, C$_{Ar}$H), 110.9 (+, C$_{Ar}$H), 34.0 (–, CH$_2$), 33.8 (–, CH$_2$), 33.3 (–, CH$_2$), 31.2 (–, CH$_2$) ppm. – ^{19}**F NMR** (101 MHz, CDCl$_3$) δ = –64.26 ppm. – **MS** (FAB, 3-NBA), *m/z* (%): 365 [M]$^+$. – **HRMS** (FAB, 3-NBA) calc. for C$_{23}$H$_{18}$N$_1$F$_3$ [M]$^+$ 365.1391, found 365.1391.

(*rac*)-1-(*N*-[2]Paracyclo[2]-6-(rifluoromethyl)(1,4)carbazolophanyl)-4-(4,6-diphenyl-1,3,5-triazin-2-yl)-benzene (133)

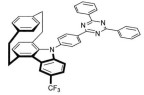

A 100 mL round bottom flask was charged with (*rac*)-[2]paracyclo[2]6-(trifluoromethyl)(1,4)carbazolophane (457 mg, 1.30 mmol, 1.00 equiv.), 2-(4-fluorophenyl)-4,6-diphenyl-1,3,5-triazine (512 mg, 1.60 mmol, 1.25 equiv.) and potassium phosphate tribasic (1.33 g, 6.30 mmol, 5.00 equiv.). It was evacuated and flushed with argon three times. Through the septum 42 mL of anhydrous DMSO were added, then it was heated to 150 °C and stirred for 12 h. After cooling to room temperature, the reaction mixture was diluted with 50 mL of dichloromethane and washed with brine (3 × 50 mL). The organic layer was dried over MgSO$_4$ and the solvent was removed under reduced pressure. The obtained crude product was purified *via* column chromatography on silica gel (cyclohexane/dichloromethane = 2.5:1) to yield the title compound as a white luminescent solid (460 mg, 0.684 mmol, 55%).

R$_f$ = 0.60 (cyclohexane/dichloromethane = 2.5:1). – ^{1}H NMR (400 MHz, CDCl$_3$) δ = 9.05 (bs, 2H), 8.91 – 8.80 (m, 4H), 8.41 (s, 1H), 8.07 (bs, 1H), 7.71–7.54 (m, 9H), 6.79–6.67 (m, 2H), 6.53 (dd, *J* = 7.9, 1.9 Hz, 1H), 6.37 (dd, *J* = 7.8, 1.8 Hz, 1H), 5.97 (dd, *J* = 7.8, 1.9 Hz, 1H), 5.52 (dd, *J* = 7.7, 1.9 Hz, 1H), 4.16–4.00 (m, 1H), 3.31–3.13 (m, 2H), 3.14–3.03 (m, 1H), 2.92 (ddd, *J* = 12.6, 8.3, 3.2 Hz, 1H), 2.82–2.70 (m, 2H), 2.38–2.23 (m, 1H) ppm. – ^{13}C NMR (101 MHz, CDCl$_3$) δ = 172.0 (C$_q$, C$_{Ar}$), 170.8 (C$_q$, C$_{Ar}$), 142.1 (C$_q$, C$_{Ar}$), 142.0 (C$_q$, C$_{Ar}$), 141.5 (C$_q$, C$_{Ar}$), 138.1 (C$_q$, C$_{Ar}$), 137.5 (C$_q$, C$_{Ar}$), 136.2 (C$_q$, C$_{Ar}$), 136.2 (C$_q$, C$_{Ar}$), 135.7 (C$_q$, C$_{Ar}$), 134.6 (+, C$_{Ar}$H), 132.9 (+, C$_{Ar}$H), 132.0 (+, C$_{Ar}$H), 131.6 (+, C$_{Ar}$H), 130.5 (+, C$_{Ar}$H), 129.2 (+, C$_{Ar}$H), 128.9 (+, C$_{Ar}$H), 128.5 (+, C$_{Ar}$H), 126.7 (+, C$_{Ar}$H), 126.5 (C$_q$, C$_{Ar}$), 125.7 (+, C$_{Ar}$H), 125.4 (C$_q$, C$_{Ar}$), 124.8 (C$_q$, C$_{Ar}$), 124.0 (C$_q$, C$_{Ar}$), 123.1 (C$_q$, q, *J* = 31.9 Hz, C$_{Ar}$). 122.3 (+, q, *J* = 3.6 Hz, C$_{Ar}$H), 119.9 (+, q, *J* = 4.1 Hz, C$_{Ar}$H), 110.1 (+, C$_{Ar}$H) ppm. – ^{19}F NMR (101 MHz, CDCl$_3$) δ = –64.41 ppm. – IR (ATR, ṽ) = 2929 (w), 1601 (w), 1588 (w), 1510 (vs), 1445 (m), 1392 (m), 1363 (vs), 1332 (vs), 1303 (s), 1268 (s), 1237 (m), 1157 (m), 1147 (m), 1129 (m), 1111 (vs), 1082 (s), 1064 (s), 1026 (m), 1014 (m), 989 (w), 887 (w), 833 (m), 800 (m), 772 (s), 764 (s), 744 (s), 737 (s), 687 (vs), 646 (s), 639 (s), 596 (m), 545 (w), 514 (s), 493 (m), 467 (w) cm^{-1}. – MS (FAB, 3-NBA), *m/z* (%): 673 [M+H]$^+$, 672 [M]$^+$. – HRMS (FAB, 3-NBA) calc. for C$_{44}$H$_{32}$N$_4$F$_3$ [M+H]$^+$ 673.2579, found 673.2580. – EA (C$_{44}$H$_{31}$F$_3$N$_4$) calc. C: 78.56, H: 4.64, N: 8.33; found C: 78.54, H: 4.71, N: 8.36.

(*rac*)-1-(*N*-[2]Paracyclo[2]-6-(bromo)(1,4)carbazolophanyl)-4-(4,6-diphenyl-1,3,5-triazin-2-yl)-benzene (145)

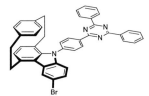

A 50 mL sealable vial was charged with (*rac*)-[2]paracyclo[2]6-(bromo)(1,4)carbazolophane (385 mg, 1.02 mmol, 1.00 equiv.), 2-(4-fluorophenyl)-4,6-diphenyl-1,3,5-triazine (401 mg, 1.22 mmol, 1.20 equiv.) and potassium phosphate tribasic (1.14 g, 5.36 mmol, 5.24 equiv.). It was evacuated and flushed with argon three times. Through the septum 25 mL of anhydrous DMSO were added, then it was heated to 120 °C and stirred for 12 h. After cooling to room temperature, the reaction mixture was diluted with 50 mL of dichloromethane and washed with brine (3 × 100 mL). The organic layer was dried over MgSO$_4$ and the solvent was removed under reduced pressure. The obtained crude product was purified *via* column chromatography on silica gel (cyclohexane/dichloromethane = 10:1 to 5:1) to yield the title compound as a yellow luminescent solid (545 mg, 0.797 mmol, 78%).

R_f = 0.50 (cyclohexane/dichloromethane = 5:1). – **^{1}H NMR** (400 MHz, CDCl$_3$) δ = 9.03 (bs, 2H), 8.90–8.70 (m, 4H), 8.26 (d, *J* = 1.8 Hz, 1H), 8.15–7.57 (m, 8H), 7.50 (dd, *J* = 8.7 Hz, *J* = 1.8 Hz, 1H), 7.41 (d, *J* = 8.7 Hz, 1H), 6.74–6.64 (m, 2H), 6.52 (dd, *J* = 7.9 Hz, *J* = 1.8 Hz, 1H), 6.35 (dd, *J* = 7.9 Hz, *J* = 1.8 Hz, 1H), 5.97 (dd, *J* = 7.8 Hz, *J* = 1.9 Hz, 1H), 5.60 (dd, *J* = 7.8 Hz, *J* = 1.9 Hz, 1H), 4.07–3.92 (m, 1H), 3.31–3.02 (m, 3H), 2.97–2.84 (m, 1H), 2.82–2.66 (m, 2H), 2.40–2.17 (m, 1H) ppm. – **^{13}C NMR** (101 MHz, CDCl$_3$) δ = 172.0 (C$_q$, C$_{Ar}$), 170.9 (C$_q$, C$_{Ar}$), 142.3 (C$_q$, C$_{Ar}$), 141.1 (C$_q$, C$_{Ar}$), 139.2 (C$_q$, C$_{Ar}$), 138.1 (C$_q$, C$_{Ar}$), 137.5 (C$_q$, C$_{Ar}$), 136.2 (C$_q$, C$_{Ar}$), 136.1 (C$_q$, C$_{Ar}$), 135.4 (C$_q$, C$_{Ar}$), 134.4 (+, C$_{Ar}$H), 132.9 (+, C$_{Ar}$H), 131.9 (+, C$_{Ar}$H), 131.5 (+, C$_{Ar}$H), 130.4 (+, C$_{Ar}$H), 129.2 (+, C$_{Ar}$H), 128.9 (+, C$_{Ar}$H), 128.1 (+, C$_{Ar}$H), 128.1 (+, C$_{Ar}$H), 127.5 (C$_q$, C$_{Ar}$), 127.0 (C$_q$, C$_{Ar}$), 126.8 (+, C$_{Ar}$H), 126.1 (+, C$_{Ar}$H), 125.8 (+, C$_{Ar}$H), 125.0 (+, C$_{Ar}$H), 124.7 (C$_q$, C$_{Ar}$), 113.8 (C$_q$, C$_{Ar}$), 111.4 (+, C$_{Ar}$H), 35.1 (–, CH$_2$), 33.8 (–, CH$_2$), 33.4 (–, CH$_2$), 33.2 (–, CH$_2$) ppm. – **MS** (FAB, 3-NBA), *m/z* (%): 682 [M]$^+$. – **HRMS** (FAB, 3-NBA) calc. for C$_{43}$H$_{31}$N$_4$^{79}Br$_1$ [M]$^+$ 682.1732, found 682.1733.

(*rac*)-1-(*N*-[2]Paracyclo[2](1,4)carbazolophanyl)-4-(4,6-diphenyl-1,3,5-triazin-2-yl)-benzene (131)

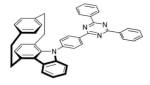

A 100 mL round bottom flask was charged with (*rac*)-[2]paracyclo[2](1,4)carbazolophane (890 mg, 2.99 mmol, 1.00 equiv.), 2-(4-fluorophenyl)-4,6-diphenyl-1,3,5-triazine (1.17 g, 3.57 mmol, 1.19 equiv.) and potassium phosphate tribasic (3.18 g, 15.0 mmol, 5.01 equiv.). It was evacuated and flushed with argon three times. Through the septum 40 mL of anhydrous DMSO were added, then it was heated to 120 °C and stirred for 12 h. After cooling to room temperature, the reaction mixture was diluted with 10 mL of dichloromethane and washed with brine (3 × 100 mL). The organic layer was dried over MgSO$_4$ and the solvent was removed under reduced pressure. The obtained crude product was purified *via* column chromatography on silica gel (cyclohexane/dichloromethane = 5:1 to 3:1) to yield the title compound as a yellow luminescent solid (1.20 g, 1.98 mmol, 66%).

R$_f$ = 0.55 (cyclohexane/dichloromethane = 2:1). – **^{1}H NMR** (400 MHz, CDCl$_3$) δ = 9.03 (bs, 2H), 8.89–8.76 (m, 4H), 8.17 (d, J = 7.6 Hz, 1H), 8.05–7.59 (m, 8H), 7.57 (d, J = 8.1 Hz, 1H), 7.43 (t, J = 7.1 Hz, 1H), 7.37 (t, J = 7.0 Hz, 1H), 6.74–6.61 (m, 2H), 6.52 (dd, J = 7.8 Hz, 1.9 Hz, 1H), 6.36 (dd, J = 7.9 Hz, J = 1.8 Hz, 1H), 5.98 (dd, J = 7.7 Hz, J = 1.9 Hz, 1H), 5.54 (dd, J = 7.7 Hz, J = 1.9 Hz, 1H), 4.25–3.89 (m, 1H), 3.29–3.01 (m, 3H), 2.98–2.85 (m, 1H), 2.85–2.65 (m, 2H), 2.30 (ddd, J = 13.0 Hz, J = 9.5 Hz, J = 7.3 Hz, 1H) ppm. – **^{13}C NMR** (101 MHz, CDCl$_3$) δ = 171.9 (C$_q$, C$_{Ar}$), 171.0 (C$_q$, C$_{Ar}$), 142.9 (C$_q$, C$_{Ar}$), 140.7 (C$_q$, C$_{Ar}$), 140.5 (C$_q$, C$_{Ar}$), 138.1 (C$_q$, C$_{Ar}$), 137.5 (C$_q$, C$_{Ar}$), 136.2 (C$_q$, C$_{Ar}$), 135.9 (C$_q$, C$_{Ar}$), 134.9 (C$_q$, C$_{Ar}$), 133.7 (+, C$_{Ar}$H), 132.8 (+, C$_{Ar}$H), 131.8 (+, C$_{Ar}$H), 131.4 (+, C$_{Ar}$H), 130.3 (+, C$_{Ar}$H), 129.1 (+, C$_{Ar}$H), 128.9 (+, C$_{Ar}$H), 127.8 (+, C$_{Ar}$H), 127.1 (C$_q$), 126.7 (+, C$_{Ar}$H), 125.9 (C$_q$, C$_{Ar}$), 125.8 (+, C$_{Ar}$H), 125.3 (+, C$_{Ar}$H), 124.6 (C$_q$, C$_{Ar}$), 122.7 (+, C$_{Ar}$H), 121.0 (+, C$_{Ar}$H), 109.9 (+, C$_{Ar}$H), 35.1 (–, CH$_2$), 34.0 (–, CH$_2$), 33.6 (–, CH$_2$), 33.2 (–, CH$_2$) ppm. – **IR** (ATR, ṽ) = 2922 (w), 1599 (w), 1588 (w), 1513 (vs), 1476 (w), 1445 (s), 1411 (w), 1394 (s), 1364 (vs), 1333 (m), 1307 (m), 1298 (m), 1254 (w), 1239 (w), 1173 (w), 1153 (w), 1146 (w), 1109 (w), 1026 (w), 1010 (w), 836 (m), 766 (vs), 739 (vs), 731 (s), 718 (m), 696 (s), 687 (s), 669 (w), 656 (m), 647 (m), 639 (w), 628 (w), 598 (w), 582 (w), 527 (m), 513 (m), 483 (w), 467 (w), 442 (w) cm^{-1}. – **MS** (FAB, 3-NBA), *m/z* (%): 605 [M+H]$^+$, 604 [M]$^+$. – **HRMS** (FAB, 3-NBA) calc. for C$_{43}$H$_{32}$N$_4$ [M]$^+$ 604.2627, found 604.2629.

Friedel-Crafts benzoylation for benzoyl-functionalized emitters (134–136)

A 100 mL round bottom flask was charged with (*rac*)-1-(*N* [2]paracyclo[2](1,4)carbazolophanyl)-4-(4,6-diphenyl-1,3,5-triazin-2-yl)-benzene (1.06 g, 1.75 mmol, 1.00 equiv.) and aluminum trichloride (520 mg, 3.90 mmol, 2.22 equiv.). It was evacuated and flushed with argon three times. Through the septum benzoyl chloride (540 mg, 3.84 mmol, 2.19 equiv.) and 45 mL of anhydrous DCM were added, then it was heated to 40 °C and stirred for 12 h. After cooling to room temperature, the reaction mixture was quenched with distilled water and diluted with 50 mL of dichloromethane. Then it was washed with brine (3 × 100 mL). The organic layer was dried over MgSO$_4$ and the solvent was removed under reduced pressure. The obtained crude product was purified *via* column chromatography on silica gel (cyclohexane/dichloromethane = 1:1 to 1:5) to yield three compounds (**134–136**) as shown below.

(*rac*)-1-(*N*-[2]Paracyclo[2]-3-(benzoyl)(1,4)carbazolophanyl)-4-(4,6-diphenyl-1,3,5-triazin-2-yl)-benzene (135)

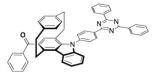

Yellow luminescent solid (710 mg, 1.00 mmol, 57%). **R**$_f$ = 0.55 (dichloromethane = 1). – **^{1}H NMR** (400 MHz, CDCl$_3$) δ = 9.30–8.90 (m, 2H), 8.90–8.78 (m, 4H), 8.26 (d, *J* = 7.9 Hz, 1H), 8.17 (bs, 1H), 7.96–7.85 (m, 2H), 7.70–7.54 (m, 9H), 7.52–7.43 (m, 3H), 7.43–7.38 (m, 1H), 7.03 (s, 1H), 6.74 (dd, *J* = 7.9 Hz, *J* = 1.9 Hz, 1H), 6.40 (dd, *J* = 8.0 Hz, *J* = 1.9 Hz, 1H), 6.01 (dd, *J* = 7.7 Hz, *J* = 1.9 Hz, 1H), 5.64 (dd, *J* = 7.8 Hz, *J* = 1.9 Hz, 1H), 3.89 (ddd, *J* = 13.3 Hz, *J* = 10.2 Hz, *J* = 3.1 Hz, 1H), 3.46 (ddd, *J* = 13.6 Hz, *J* = 10.5 Hz, *J* = 4.7 Hz, 1H), 3.25 (ddd, *J* = 13.5 Hz, *J* = 10.5 Hz, *J* = 3.0 Hz, 1H), 3.08 (ddd, *J* = 12.9 Hz, *J* = 10.1 Hz, *J* = 4.7 Hz, 1H), 2.98–2.86 (m, 1H), 2.83–2.62 (m, 2H), 2.28 (ddd, *J* = 13.2 Hz, *J* = 9.4 Hz, *J* = 7.6 Hz, 1H) ppm. – **^{13}C NMR** (101 MHz, CDCl$_3$) δ = 197.0 (C$_q$, CO), 172.0 (C$_q$, C$_{Ar}$), 170.9 (C$_q$, C$_{Ar}$), 142.1 (C$_q$, C$_{Ar}$), 142.0 (C$_q$, C$_{Ar}$), 141.4 (C$_q$, C$_{Ar}$), 140.2 (C$_q$, C$_{Ar}$), 138.9 (C$_q$, C$_{Ar}$), 137.9 (C$_q$, C$_{Ar}$), 137.8 (C$_q$, C$_{Ar}$), 136.2 (C$_q$, C$_{Ar}$), 135.8 (+, C$_{Ar}$H), 135.6 (C$_q$, C$_{Ar}$), 133.4 (C$_q$, C$_{Ar}$), 132.9 (+, C$_{Ar}$H), 132.5 (+, C$_{Ar}$H), 131.1 (+, C$_{Ar}$H), 130.7 (+, C$_{Ar}$H), 130.4 (+, C$_{Ar}$H), 130.3 (+, C$_{Ar}$H), 129.2 (+, C$_{Ar}$H), 128.9 (+, C$_{Ar}$H), 128.4 (+, C$_{Ar}$H), 127.5 (C$_q$, C$_{Ar}$), 126.7 (+, C$_{Ar}$H), 126.0 (+, C$_{Ar}$H), 126.0 (+, C$_{Ar}$H), 125.7 (C$_q$, C$_{Ar}$), 123.1 (+, C$_{Ar}$H), 121.6 (+, C$_{Ar}$H), 110.3 (+, C$_{Ar}$H), 34.9 (–, CH$_2$), 33.7 (–, CH$_2$), 33.7(–, CH$_2$), 30.7(–, CH$_2$) ppm. – **IR** (ATR, ṽ) = 2921 (w), 2847 (w), 1646 (w), 1588 (w), 1510 (vs), 1477 (w), 1441 (s), 1411 (w), 1398 (w), 1364 (vs), 1337 (s),

1315 (m), 1298 (m), 1244 (s), 1231 (m), 1173 (m), 1157 (w), 1147 (w), 1091 (w), 1068 (w), 1026 (w), 1001 (w), 898 (w), 833 (m), 793 (w), 765 (s), 739 (s), 688 (vs), 670 (m), 662 (m), 645 (m), 637 (m), 618 (w), 602 (w), 582 (w), 516 (m), 486 (w), 469 (w), 463 (w), 436 (w) cm^{-1}. – **MS** (FAB, 3-NBA), m/z (%): 709 [M+H]$^+$, 708 [M]$^+$. – **HRMS** (FAB, 3-NBA) calc. for $C_{50}H_{37}O_1N_4$ [M+H]$^+$ 709.2967, found 709.2965. – **EA** ($C_{50}H_{36}N_4O$) calc. C: 84.72, H: 5.12, N: 7.90; found C: 82.60, H: 5.64, N: 7.04.

(*rac*)-1-(*N*-[2]Paracyclo[2]-6-(benzoyl)(1,4)carbazolophanyl)-4-(4,6-diphenyl-1,3,5-triazin-2-yl)-benzene (134)

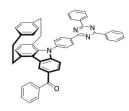

Yellow luminescent solid (35.0 mg, 49.4 µmol, 3%). R_f = 0.50 (dichloromethane = 1). – **^{1}H NMR** (400 MHz, CDCl$_3$) δ = 9.06 (bs, 2H), 8.89–8.76 (m, 4H), 8.66 (d, J = 1.4 Hz, 1H), 8.25–7.88 (m, 4H), 7.76–7.49 (m, 11H), 6.83–6.65 (m, 2H), 6.53 (dd, J = 7.9 Hz, 1.9 Hz, 1H), 6.37 (dd, J = 7.9 Hz, J = 1.9 Hz, 1H), 6.01 (dd, J = 7.7 Hz, J = 2.0 Hz, 1H), 5.62 (dd, J = 7.7 Hz, J = 1.9 Hz, 1H), 4.09–3.92 (m, 1H), 3.28–3.02 (m, 3H), 2.92 (ddd, J = 12.5 Hz, J = 8.7 Hz, J = 3.0 Hz, 1H), 2.83–2.67 (m, 2H), 2.40–2.23 (m, 1H) ppm.

(*rac*)-1-(*N*-[2]Paracyclo[2]-3,6-di(benzoyl)(1,4)carbazolophanyl)-4-(4,6-diphenyl-1,3,5-triazin-2-yl)-benzene (136)

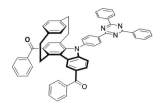

Yellow luminescent solid (340 mg, 418 µmol, 24%). R_f = 0.20 (dichloromethane = 1). – **^{1}H NMR** (400 MHz, CDCl$_3$) δ = 9.10 (d, J = 81.5 Hz, 2H), 8.90–8.80 (m, 4H), 8.76 (d, J = 1.6 Hz, 1H), 8.19 (bs, 1H), 7.99 (dd, J = 8.6 Hz, J = 1.6 Hz, 1H), 7.97–7.86 (m, 4H), 7.70–7.52 (m, 12H), 7.52–7.43 (m, 2H), 7.08 (s, 1H), 6.76 (dd, J = 8.0 Hz, J = 1.9 Hz, 1H), 6.43 (dd, J = 8.0 Hz, J = 1.9 Hz, 1H), 6.05 (dd, J = 7.8 Hz, J = 1.9 Hz, 1H), 5.71 (dd, J = 7.8 Hz, J = 1.9 Hz, 1H), 3.80 (ddd, J = 13.3 Hz, J = 10.2 Hz, J = 2.9 Hz, 1H), 3.40 (ddd, J = 13.9 Hz, J = 10.5 Hz, J = 4.7 Hz, 1H), 3.25 (ddd, J = 13.4 Hz, J = 10.5 Hz, J = 2.8 Hz, 1H), 3.07 (ddd, J = 13.3 Hz, J = 10.3 Hz, J = 4.7 Hz, 1H), 3.00–2.89 (m, 1H), 2.82–2.67 (m, 2H), 2.38–2.22 (m, 1H) ppm. – **^{13}C NMR** (101

MHz, CDCl$_3$) δ = 196.8 (C$_q$, C$_{Ar}$), 196.6 (C$_q$, C$_{Ar}$), 172.1 (C$_q$, C$_{Ar}$), 170.7 (C$_q$, C$_{Ar}$), 143.8 (C$_q$, C$_{Ar}$), 143.0 (C$_q$, C$_{Ar}$), 141.3(C$_q$, C$_{Ar}$), 139.8 (C$_q$, C$_{Ar}$), 138.8 (C$_q$, C$_{Ar}$), 138.7 (C$_q$, C$_{Ar}$), 138.0 (C$_q$, C$_{Ar}$), 137.8 (C$_q$, C$_{Ar}$), 136.3 (C$_q$, C$_{Ar}$), 136.2 (+, C$_{Ar}$H), 136.1 (C$_q$, C$_{Ar}$), 134.3 (C$_q$, C$_{Ar}$), 133.0 (+, C$_{Ar}$H), 132.7 (+, C$_{Ar}$H), 132.3 (+, C$_{Ar}$H), 131.4 (+, C$_{Ar}$H), 131.1 (+, C$_{Ar}$H), 130.6 (+, C$_{Ar}$H), 130.2 (+, C$_{Ar}$H), 129.2 (+, C$_{Ar}$H), 128.9 (+, C$_{Ar}$H), 128.6 (+, C$_{Ar}$H), 128.5 (+, C$_{Ar}$H), 128.4 (+, C$_{Ar}$H), 127.4 (C$_q$, C$_{Ar}$), 126.6 (+, C$_{Ar}$H), 126.4 (+, C$_{Ar}$H), 126.0 (+, C$_{Ar}$H), 125.3 (C$_q$, C$_{Ar}$), 123.6 (C$_q$, C$_{Ar}$), 109.9 (+, C$_{Ar}$H), 34.9 (–, CH$_2$), 33.7 (–, CH$_2$), 33.6 (–, CH$_2$), 30.8 (–, CH$_2$) ppm. – **IR** (ATR, ṽ) = 2921 (w), 2850 (w), 1649 (s), 1588 (m), 1513 (vs), 1443 (s), 1364 (vs), 1341 (s), 1329 (s), 1315 (s), 1307 (s), 1268 (vs), 1248 (vs), 1173 (m), 1133 (m), 899 (m), 829 (m), 768 (s), 737 (vs), 710 (vs), 703 (vs), 690 (vs), 645 (s), 637 (s), 524 (m) cm^{-1}. – **MS** (FAB, 3-NBA), *m/z* (%): 813 [M+H]$^+$, 812 [M]$^+$. – **HRMS** (FAB, 3-NBA) calc. for C$_{57}$H$_{41}$O$_2$N$_4$ [M+H]$^+$ 813.3230, found 813.3230.

4, 16-*N*-Di(2-chlorophenyl)amino[2.2]paracyclophane (152)

 Under argon atmosphere, a mixture of 4,16-dibromo[2.2]paracyclophane (1.10 g, 3.00 mmol, 1.00 equiv.), 2-chloroaniline (803 mg, 6.29 mmol, 2.10 equiv.), Pd(OAc)₂ (68.0 mg, 303 µmol, 10 mol%), tri-*tert*-butylphosphonium tetrafluoroborate (86.0 mg, 296 µmol, 10 mol%) and sodium *tert*-butoxide (717 mg, 7.46 mmol, 2.49 equiv.) in toluene (36.0 mL) was stirred at 100 °C for 12 h. The mixture was cooled to room temperature and diluted with dichloromethane (100 mL). Then it was washed with brine (3 × 100 mL). The organic layer was dried over MgSO₄ and the solvent was removed under reduced pressure. The obtained crude product was purified *via* column chromatography on silica gel (cyclohexane/dichloromethane = 5:1 to 3:1) to yield the title compound as a yellow solid (1.04 g, 2.27 mmol, 76%).

R_f = 0.30 (cyclohexane/dichloromethane = 5:1). – **¹H NMR** (400 MHz, CDCl₃) δ = 7.37 (dd, J = 7.9 Hz, 1.5 Hz, 2H), 7.07–7.01 (m, 2H), 6.95 (dd, J = 7.8 Hz, J = 1.8 Hz, 2H), 6.87 (dd, J = 8.2 Hz, J = 1.5 Hz, 2H), 6.80–6.70 (m, 2H), 6.50 (d, J = 7.8 Hz, 2H), 5.99 (s, 2H), 5.96 (d, J = 1.8 Hz, 2H), 3.06 (ddd, J = 13.5 Hz, J = 8.2, J = 3.5 Hz, 2H), 3.01–2.88 (m, 4H), 2.68 (ddd, J = 13.5 Hz, J = 9.7 Hz, J = 6.9 Hz, 2H) ppm. – **¹³C NMR** (101 MHz, CDCl₃) δ = 141.0 (C_q), 140.4 (C_q), 138.9 (C_q), 134.2 (+, C_ArH), 132.8 (C_q), 129.5 (+, C_ArH), 128.4 (+, C_ArH), 127.6 (+, C_ArH), 125.0 (+, C_ArH), 120.2 (C_q), 119.6 (+, C_ArH), 114.2 (+, C_ArH), 33.3 (–, CH₂), 32.9 (–, CH₂) ppm. – **IR** (ATR, ṽ) = 3400 (m), 2941 (w), 2924 (w), 2851 (w), 1587 (s), 1558 (w), 1504 (s), 1487 (vs), 1455 (m), 1435 (vs), 1310 (s), 1286 (s), 1256 (m), 1224 (m), 1194 (w), 1176 (w), 1153 (m), 1140 (w), 1123 (w), 1094 (w), 1048 (m), 1030 (vs), 987 (w), 967 (w), 948 (w), 925 (w), 908 (w), 888 (s), 860 (w), 840 (w), 799 (m), 761 (w), 734 (vs), 711 (s), 686 (s), 657 (m), 629 (w), 586 (m), 535 (w), 526 (w), 499 (s), 470 (w), 453 (m), 433 (s), 401 (m), 397 (m), 381 (s) cm⁻¹. – **MS** (EI, 180 °C), *m/z* (%): 458 [M]⁺. – **HRMS** (EI, 180 °C) calc. for C₂₈H₂₄O₂N₂³⁵Cl₂ [M]⁺ 458.1317, found 458.1318.

trans-anti-[2.2]-1,4-1',4'-dicarbazolocyclophane (148, dicarbazolocyclophane, DCCP)

Under argon atmosphere, a mixture of 4,16-N-di(2-chlorophenyl)amino[2.2]paracyclophane (4.59 g, 9.99 mmol, 1.00 equiv.), Pd$_2$(dba)$_3$ (1.83 g, 2.00 mmol, 20 mol%), 2-dicyclohexylphosphino-2',4',6'-triisopropylbiphenyl (XPhos, 2.86 g, 6.00 mmol, 60 mol%), pivalic acid (1.23 g, 12.0 mmol, 1.21 equiv.) and potassium carbonate (13.8 g, 99.9 mmol, 9.99 equiv.) in anhydrous N,N-dimethylacetamide (60 mL) was stirred at 110 °C for 12 h. The mixture was cooled to room temperature and diluted with DCM (200 mL). Then it was filtered over a short silica gel plate and thoroughly washed with DCM. The filtrate was washed with brine (3 × 200 mL) and the organic solvent was removed under reduced pressure. The final product was obtained through recrystallization from THF as a pale grey precipitate (1.89 g, 4.88 mmol, 49%).

R_f = 0.40 (cyclohexane/ dichloromethane = 1:1). – **^{1}H NMR** (400 MHz, DMSO-d_6) δ = 10.98 (s, 2H), 8.06 (d, J = 7.8 Hz, 2H), 7.51 (d, J = 8.0 Hz, 2H), 7.43–7.31 (m, 2H), 7.21–7.15 (m, 2H), 5.51 (d, J = 7.4 Hz, 2H), 5.03 (d, J = 7.4 Hz, 2H), 3.83 (dd, J = 13.2 Hz, J = 10.0 Hz, 2H), 3.52 (dd, J = 13.3 Hz, J = 9.9 Hz, 2H), 3.13 (ddd, J = 13.2 Hz, J = 10.0 Hz, J = 6.8 Hz, 2H), 2.90 (ddd, J = 13.2 Hz, J = 10.1 Hz, J = 6.8 Hz, 2H) ppm. – **^{13}C NMR** (101 MHz, DMSO-d_6) δ = 139.8 (C$_q$, C$_{Ar}$), 139.5 (C$_q$, C$_{Ar}$), 133.2 (C$_q$, C$_{Ar}$), 125.3 (+, C$_{Ar}$H), 124.4 (C$_q$, C$_{Ar}$), 124.1 (+, C$_{Ar}$H), 123.3(C$_q$, C$_{Ar}$), 121.5 (+, C$_{Ar}$H), 120.4 (+, C$_{Ar}$H), 119.8 (C$_q$, C$_{Ar}$), 118.3 (+, C$_{Ar}$H), 110.7 (+, C$_{Ar}$H), 31.2 (–, CH$_2$), 29.3 (–, CH$_2$) ppm. – **IR** (ATR, $\tilde{\nu}$) = 3403 (m), 3040 (w), 2929 (w), 2859 (w), 1595 (w), 1565 (w), 1504 (w), 1453 (s), 1441 (w), 1395 (m), 1390 (m), 1320 (s), 1299 (w), 1288 (w), 1266 (m), 1237 (w), 1116 (w), 1027 (m), 926 (w), 857 (w), 824 (w), 771 (s), 749 (vs), 735 (vs), 711 (w), 696 (w), 676 (m), 596 (s), 574 (w), 561 (m), 533 (s), 448 (s), 436 (s), 424 (m), 415 (m), 399 (s), 391 (s), 385 (s), 378 (s) cm^{-1}. – **MS** (FAB, 3-NBA), m/z (%): 386 [M]$^+$. – **HRMS** (FAB, 3-NBA) calc. for C$_{28}$H$_{22}$N$_2$ [M]$^+$ 386.1783, found 386.1784.

trans-anti-9,9'-Bis(4-(4,6-bis(4-(tert-butyl)phenyl)-1,3,5-triazin-2-yl)phenyl)[2.2]-1,4-1',4'-dicarbazolocyclophane (149)

A 20 mL sealable vial was charged with _trans-anti_-[2.2]-1,4-1',4'-dicarbazoloc-yclophane (58.0 mg, 150 μmol, 1.00 equiv.), 2,4-bis(4-(tert-butyl)phenyl)-6-(4-fluorophenyl)-1,3,5-triazine (139 mg, 316 μmol, 2.11 equiv.) and potassium phosphate tribasic (320 mg, 1.51 mmol, 10.00 equiv.). It was evacuated and flushed with argon three times. Through the septum 4 mL of anhydrous DMSO were added, then it was heated to 120 °C and stirred for 12 h. After cooling to room temperature, the reaction mixture was diluted with 50 mL of dichloromethane and washed with brine (3 × 50 mL). The organic layer was dried over MgSO$_4$ and the solvent was removed under reduced pressure. The obtained crude product was purified _via_ column chromatography on silica gel (cyclohexane/dichloromethane = 3:1 to 1.5:1) to yield the title compound as a yellow luminescent solid (140 mg, 114 μmol, 76%).

R$_f$ = 0.40 (cyclohexane/dichloromethane = 2:1). – **^{1}H NMR** (400 MHz, CDCl$_3$) δ = 9.05 (d, J = 87.6 Hz, 4H), 8.85–8.65 (m, 8H), 8.24 (bs, 2H), 8.04 (d, J = 7.8 Hz, 2H), 7.65 (d, J = 8.6 Hz, 8H), 7.59 (d, J = 8.2 Hz, 2H), 7.51–7.38 (m, 4H), 7.33 (t, J = 7.2 Hz, 2H), 6.04 (d, J = 7.6 Hz, 2H), 5.65 (d, J = 7.6 Hz, 2H), 3.74 (dd, J = 13.4 Hz, J = 9.3 Hz, 2H), 3.04 (dt, J = 13.7 Hz, J = 9.1 Hz, 2H), 2.87 (dd, J = 13.8 Hz, J = 9.2 Hz, 2H), 2.57 (dt, J = 13.6 Hz, J = 9.1 Hz, 2H), 1.44 (s, 36H) ppm. – **^{13}C NMR** (101 MHz, CDCl$_3$) δ = 171.8 (C$_q$, C$_{Ar}$), 170.8 (C$_q$, C$_{Ar}$), 156.4 (C$_q$, C$_{Ar}$), 142.7 (C$_q$, C$_{Ar}$), 140.8 (C$_q$, C$_{Ar}$), 140.3 (C$_q$, C$_{Ar}$), 135.2 (C$_q$, C$_{Ar}$), 134.4 (C$_q$, C$_{Ar}$), 133.7 (C$_q$, C$_{Ar}$), 130.2 (+, C$_{Ar}$H), 129.0 (+, C$_{Ar}$H), 125.9 (+, C$_{Ar}$H), 125.7 (C$_q$, C$_{Ar}$), 125.5 (C$_q$, C$_{Ar}$), 125.2 (+, C$_{Ar}$H), 123.1 (+, C$_{Ar}$H), 122.3 (+, C$_{Ar}$H), 120.9 (+, C$_{Ar}$H), 109.9 (+, C$_{Ar}$H), 35.3 (C$_q$, CCH$_3$), 33.6 (–, CH$_2$), 31.4 (+, CH$_3$) ppm. – **IR** (ATR, ṽ) = 2958 (w), 2952 (w), 1578 (m), 1504 (vs), 1477 (m), 1452 (s), 1409 (m), 1395 (s), 1367 (vs), 1333 (m), 1320 (m), 1310 (m), 1264 (s), 1234 (w), 1190 (m), 1170 (w), 1150 (w), 1105 (w), 1016 (m), 854 (w), 815 (vs), 769 (m), 739 (s), 727 (s), 704 (m), 688 (w), 589 (w), 551 (s), 533 (w), 482 (w), 462 (w) cm^{-1}. – **MS** (FAB, 3-NBA), _m/z_ (%): 1225 [M]$^+$.

trans-anti-6,6'-Dibromo[2.2]-1,4-1',4'-dicarbazolocyclophane (156)

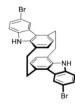

A solution of NBS (156 mg, 876 µmol, 2.20 equiv.) in 10.0 mL of THF was added slowly to a solution of *trans-anti*-[2.2]-1,4-1',4'-dicarbazolocyclophane (154 mg, 398 µmol, 1.00 equiv.) in 30.0 mL of THF in a round bottom flask at 0 °C. The reaction mixture was stirred at this temperature for 1 h. Then the solvent was removed under reduced pressure. The obtained mixture was thoroughly washed with methanol to yield the final product as a pale grey solid (140 mg, 257 µmol, 65%).

R_f = 0.35 (cyclohexane/dichloromethane = 1:1). – **^{1}H NMR** (400 MHz, DMSO-d_6) δ = 11.24 (s, 2H), 8.18 (s, 2H), 7.48 (d, J = 1.2 Hz, 4H), 5.56 (d, J = 7.4 Hz, 2H), 5.12 (d, J = 7.4 Hz, 2H), 4.21–4.01 (m, 2H), 3.92–3.75 (m, 2H), 3.61–3.44 (m, 2H), 2.98–2.82 (m, 2H) ppm. – **IR** (ATR, ṽ) = 3403 (s), 2932 (w), 2915 (w), 2859 (w), 1844 (vw), 1711 (vw), 1592 (w), 1506 (w), 1452 (vs), 1436 (m), 1392 (w), 1381 (w), 1300 (s), 1286 (s), 1269 (m), 1262 (m), 1239 (m), 1213 (w), 1129 (w), 1072 (w), 1021 (m), 924 (m), 856 (s), 793 (vs), 766 (s), 724 (w), 704 (w), 690 (m), 650 (w), 605 (s), 578 (m), 568 (vs), 535 (s), 460 (vs), 428 (vs), 404 (s), 378 (w) cm^{-1}. – **MS** (FAB, 3-NBA), m/z (%): 544 [M+H]$^+$, 543 [M]$^+$. – **HRMS** (FAB, 3-NBA) calc. for $C_{28}H_{20}N_2{}^{79}Br_2$ [M]$^+$ 541.9993, found 541.9994; calc. for $C_{28}H_{20}N_2{}^{79}Br_1{}^{81}Br_1$ [M]$^+$ 543.9973, found 543.9973.

trans-anti-9,9'-Bis(4-(4,6-bis(3,5-di-tert-butylphenyl)-1,3,5-triazin-2-yl)phenyl)-[2.2]-1,4-1',4'-dicarbazolocyclophane (153)

A 20 mL sealable vial was charged with _trans-anti_-[2.2]-1,4-1',4'-dicarbazolocyclophane (38.0 mg, 98.3 μmol, 1.00 equiv.), 2,4-bis(3,5-di-_tert_-butylphenyl)-6-(4-fluorophenyl)-1,3,5-triazine (116 mg, 210 μmol, 2.14 equiv.) and potassium phosphate tribasic (212 mg, 999 μmol, 10.20 equiv.). It was evacuated and flushed with argon three times. Through the septum 4 mL of anhydrous DMSO were added, then it was heated to 120 °C and stirred for 12 h. After cooling to room temperature, the reaction mixture was diluted with 50 mL of dichloromethane and washed with brine (3 × 50 mL). The organic layer was dried over MgSO$_4$ and the solvent was removed under reduced pressure. The obtained crude product was purified _via_ column chromatography on silica gel (cyclohexane/dichloromethane = 10:1 to 5:1) to yield the title compound as a yellow luminescent solid (90.0 mg, 62.1 μmol, 63%).

R$_f$ = 0.30 (cyclohexane/ethyl acetate = 50:1). – **^{1}H NMR** (400 MHz, CDCl$_3$) δ = 9.06 (d, _J_ = 91.2 Hz, 4H), 8.73 (d, _J_ = 1.9 Hz, 8H), 8.28 (bs, 2H), 8.05 (d, _J_ = 7.8 Hz, 2H), 7.75 (t, _J_ = 1.9 Hz, 4H), 7.61 (d, _J_ = 8.2 Hz, 2H), 7.56–7.38 (m, 4H), 7.34 (t, _J_ = 7.4 Hz, 2H), 6.04 (d, _J_ = 7.5 Hz, 2H), 5.65 (d, _J_ = 7.6 Hz, 2H), 3.75 (dd, _J_ = 13.4 Hz, _J_ = 9.2 Hz, 2H), 3.06 (dt, _J_ = 13.8 Hz, _J_ = 9.0 Hz, 2H), 2.91 (dd, _J_ = 13.8 Hz, _J_ = 9.1 Hz, 2H), 2.59 (dt, _J_ = 13.2 Hz, _J_ = 9.0 Hz, 2H), 1.50 (s, 72H) ppm. – **^{13}C NMR** (101 MHz, CDCl$_3$) δ = 172.4 (C$_q$, C$_{Ar}$), 170.9 (C$_q$, C$_{Ar}$), 151.4 (C$_q$, C$_{Ar}$), 142.7 (C$_q$, C$_{Ar}$), 140.8 (C$_q$, C$_{Ar}$), 140.3 (C$_q$, C$_{Ar}$), 135.6 (C$_q$, C$_{Ar}$), 135.1 (C$_q$, C$_{Ar}$), 134.4 (C$_q$, C$_{Ar}$), 130.2 (+, C$_{Ar}$H), 129.1 (+, C$_{Ar}$H), 127.2 (+, C$_{Ar}$H), 125.7 (C$_q$, C$_{Ar}$), 125.5 (C$_q$, C$_{Ar}$), 125.2 (+, C$_{Ar}$H), 123.3 (+, C$_{Ar}$H), 123.1 (+, C$_{Ar}$H), 122.3 (+, C$_{Ar}$H), 120.9 (+, C$_{Ar}$H), 109.9 (+, C$_{Ar}$H), 35.3 (C$_q$, _C_CH$_3$), 33.6 (–, CH$_2$), 31.6 (+, _C_H$_3$), 31.5 (–, CH$_2$) ppm. – **IR** (ATR, ṽ) = 2952 (s), 2902 (m), 1575 (w), 1513 (vs), 1480 (vs), 1453 (vs), 1395 (vs), 1363 (vs), 1323 (s), 1293 (s), 1259 (vs), 1249 (s), 1232 (vs), 1075 (s), 1034 (s), 962 (m), 891 (m), 878 (s), 827 (m), 816 (vs), 805 (vs), 741 (s), 707 (m), 690 (s), 649 (m), 615 (s), 545 (m), 528 (m), 470 (m), 452 (s), 425 (m), 418 (m) cm^{-1}. – **MS** (FAB, 3-NBA), _m/z_ (%): 1449 [M]$^+$.

trans-anti-9,9'-Bis(4-(4,6-bis(3,5-di-tert-butylphenyl)-1,3,5-triazin-2-yl)phenyl)-6,6'-di(bromo)[2.2]-1,4-1',4'-dicarbazolocyclophane (157)

A 20 mL sealable vial was charged with *trans-anti*-6,6'-dibromo[2.2]-1,4-1',4'-dicarbazolocyclophane (109 mg, 200 μmol, 1.00 equiv.), 2,4-bis(3,5-di-*tert*-butylphenyl)-6-(4-fluorophenyl)-1,3,5-triazine (276 mg, 500 μmol, 2.50 equiv.) and potassium phosphate tribasic (424 mg, 2.00 mmol, 10.00 equiv.). It was evacuated and flushed with argon three times. Through the septum 4 mL of anhydrous DMSO were added, then it was heated to 150 °C and stirred for 12 h. After cooling to room temperature, the reaction mixture was diluted with 50 mL of dichloromethane and washed with brine (3 × 50 mL). The organic layer was dried over MgSO$_4$ and the solvent was removed under reduced pressure. The obtained crude product was purified *via* column chromatography on silica gel (cyclohexane/dichloromethane = 10:1 to 5:1) to yield the title compound as a yellow luminescent solid (196 mg, 122 μmol, 61%).

R$_f$ = 0.70 (cyclohexane/dichloromethane = 3:1). – **^{1}H NMR** (400 MHz, CDCl$_3$) δ = 9.07 (d, J = 96.0 Hz, 4H), 8.74 (d, J = 1.9 Hz, 8H), 8.23 (bs, 2H), 8.14 (d, J = 1.7 Hz, 2H), 7.76 (t, J = 1.9 Hz, 4H), 7.57–7.43 (m, 6H), 6.05 (d, J = 7.6 Hz, 2H), 5.74 (d, J = 7.6 Hz, 2H), 3.69 (dd, J = 13.5 Hz, J = 9.3 Hz, 2H), 3.05 (dt, J = 13.8 Hz, J = 9.1 Hz, 2H), 2.92 (dd, J = 14.0 Hz, J = 9.2 Hz, 2H), 2.58 (dt, J = 13.6 Hz, J = 9.5 Hz, 2H), 1.51 (s, 72H) ppm. – **^{13}C NMR** (101 MHz, CDCl$_3$) δ = 172.4 (C$_q$, C$_{Ar}$), 170.7 (C$_q$, C$_{Ar}$), 151.4 (C$_q$, C$_{Ar}$), 142.0 (C$_q$, C$_{Ar}$), 140.7 (C$_q$, C$_{Ar}$), 139.4 (C$_q$, C$_{Ar}$), 135.7 (C$_q$, C$_{Ar}$), 135.6 (C$_q$, C$_{Ar}$), 134.5 (C$_q$, C$_{Ar}$), 130.4 (+, C$_{Ar}$H), 129.6 (+, C$_{Ar}$H), 128.1 (+, C$_{Ar}$H), 127.3 (C$_q$, C$_{Ar}$), 127.2 (+, C$_{Ar}$H), 124.7 (+, C$_{Ar}$H), 124.5 (C$_q$), 123.3 (+, C$_{Ar}$H), 123.3 (+, C$_{Ar}$H), 122.4 (C$_q$, C$_{Ar}$), 113.7 (C$_q$, C$_{Ar}$), 111.5 (+, C$_{Ar}$H), 35.3 (C$_q$, CCH$_3$), 33.5 (–, CH$_2$), 31.6 (+, CH$_3$), 31.3 (–, CH$_2$) ppm. – **IR** (ATR, ṽ) = 3403 (w), 2961 (m), 2953 (m), 1585 (m), 1509 (vs), 1449 (s), 1435 (vs), 1390 (m), 1364 (vs), 1356 (vs), 1310 (s), 1282 (s), 1269 (s), 1247 (m), 1225 (w), 1164 (w), 1154 (m), 1030 (m), 888 (m), 829 (m), 796 (m), 735 (s), 707 (m), 697 (m), 688 (vs), 659 (m), 499 (m), 479 (w), 433 (m), 412 (w), 404 (w), 390 (w), 381 (m) cm^{-1}. – **MS** (FAB, 3-NBA), *m/z* (%): 1607 [M]$^+$.

trans-anti-9,9'-Bis(4-(4,6-bis(3,5-di-tert-butylphenyl)-1,3,5-triazin-2-yl)phenyl)-6,6'-bis(3,6-di-tert-butyl-9N-carbazolyl)[2.2]-1,4-1',4'-dicarbazolocyclophane (154)

Under argon atmosphere, a mixture of *trans-anti*-9,9'-bis(4-(4,6-bis(3,5-di-tert-butylphenyl)-1,3,5-triazin-2-yl)phenyl)-6,6'-di(bromo)-[2.2]-1,4-1',4'-dicarbazolocyclophane (80.4 mg, 50.0 μmol, 1.00 equiv.), 3,6-ditert-butyl-9N-carbazole (30.7 mg, 110 μmol, 2.20 equiv.), Pd$_2$(dba)$_3$ (4.60 mg, 5.02 μmol, 10 mol%), tri-*tert*-butylphosphonium tetrafluoroborate (2.9 mg, 10.00 μmol, 20 mol%) and sodium *tert*-butoxide (48.1 mg, 501 μmol, 10.00 equiv.) in toluene (6.0 mL) was stirred at 100 °C for 12 h. The mixture was cooled to room temperature and diluted with dichloromethane (50 mL). Then it was washed with brine (3 × 50 mL). The organic layer was dried over MgSO$_4$ and the solvent was removed under reduced pressure. The obtained crude product was purified *via* column chromatography on silica gel (cyclohexane/dichloromethane = 10:1 to 5:1) to yield the title compound as a yellow luminescent solid (75.0 mg, 37.4 μmol, 75%).

R$_f$ = 0.60 (cyclohexane/ethyl acetate = 50:1). – **^{1}H NMR** (400 MHz, CDCl$_3$) δ = 9.12 (d, J = 88.4 Hz, 4H), 8.74 (d, J = 1.9 Hz, 8H), 8.37 (bs, 2H), 8.21 (d, J = 1.9 Hz, 2H), 8.20 (d, J = 1.9 Hz, 4H), 7.81 (d, J = 8.6 Hz, 2H), 7.75 (t, J = 1.9 Hz, 4H), 7.64–7.32 (m, 12H), 6.23 (d, J = 7.6 Hz, 2H), 5.98 (d, J = 7.6 Hz, 2H), 3.68 (dd, J = 13.4 Hz, J = 9.2 Hz, 2H), 3.16 (dt, J = 13.6 Hz, J = 9.0 Hz, 2H), 3.01 (dd, J = 13.8 Hz, J = 9.2 Hz, 2H), 2.64 (dt, J = 13.6 Hz, J = 9.1 Hz, 2H), 1.50 (s, 72H), 1.49 (s, 36H) ppm. – **^{13}C NMR** (101 MHz, CDCl$_3$) δ = 172.4 (C$_q$, C$_{Ar}$), 170.7 (C$_q$, C$_{Ar}$), 151.4 (C$_q$, C$_{Ar}$), 142.6 (C$_q$, C$_{Ar}$), 142.2 (C$_q$, C$_{Ar}$), 141.0 (C$_q$, C$_{Ar}$), 140.4(C$_q$, C$_{Ar}$), 139.7 (C$_q$, C$_{Ar}$), 135.7 (C$_q$, C$_{Ar}$), 135.6 (C$_q$, C$_{Ar}$), 134.7 (C$_q$, C$_{Ar}$), 131.2 (C$_q$, C$_{Ar}$), 130.4 (+, C$_{Ar}$H), 129.5 (+, C$_{Ar}$H), 127.2 (+, C$_{Ar}$H), 126.5 (C$_q$, C$_{Ar}$), 125.3 (C$_q$, C$_{Ar}$), 124.9 (+, C$_{Ar}$H), 123.8 (+, C$_{Ar}$H), 123.4 (+, C$_{Ar}$H), 123.3 (+, C$_{Ar}$H), 123.2 (C$_q$, C$_{Ar}$), 122.6 (C$_q$, C$_{Ar}$), 121.2 (+, C$_{Ar}$H), 116.4 (+, C$_{Ar}$H), 111.0 (+, C$_{Ar}$H), 109.1 (+, C$_{Ar}$H), 35.2 (C$_q$, CCH$_3$), 34.9 (C$_q$, CCH$_3$), 33.6 (–, CH$_2$), 32.2 (+, CH$_3$), 31.6 (+, CH$_3$), 30.3 (–, CH$_2$) ppm. – **IR** (ATR, ṽ) = 2953 (m), 2904 (w), 1514 (vs), 1479 (vs), 1463 (m), 1451 (s), 1438 (m), 1392 (m), 1364 (vs), 1319 (w), 1296 (m), 1281 (m), 1264 (s), 1247 (m), 1164 (w), 891 (w), 829 (m), 806 (m), 741 (w), 698 (m), 690 (m), 657 (w), 613 (w) cm^{-1}. – **MS** (FAB, 3-NBA), m/z (%): 2004 [M]$^+$.

3,6-dimethoxy-9H-carbazole (159)

A 100 mL round flask was charged with 3,6-dibromo-9H-carbazole (5.00 g, 15.4 mmol, 1.00 equiv.) and copper(I) iodine (11.8 g, 61.7 mmol, 4.01 equiv.). It was evacuated and flushed with argon three times. Through the septum 10 mL of anhydrous DMSO and 35 mL of MeONa/MeOH (5.4 M) solution were added, then it was heated to 120 °C and stirred for 3 h. After cooling to room temperature, the reaction mixture was diluted with 50 mL of dichloromethane and washed with brine (3 × 50 mL). The organic layer was dried over MgSO$_4$ and the solvent was removed under reduced pressure. The obtained crude product was purified *via* column chromatography on silica gel (cyclohexane/dichloromethane = 1:1) to yield the title compound as a white solid (2.50 g, 11.0 mmol, 72%).

R_f = 0.15 (cyclohexane/dichloromethane = 1:1). – **^{1}H NMR** (400 MHz, DMSO-d_6) δ = 10.81 (s, 1H), 7.68 (d, J = 2.6 Hz, 2H), 7.35 (d, J = 8.7 Hz, 2H), 7.00 (dd, J = 8.8, 2.5 Hz, 2H), 3.85 (s, 6H) ppm. – **^{13}C NMR** (101 MHz, DMSO-d_6) δ = 152.6 (C$_q$, C$_{Ar}$), 135.3 (C$_q$, C$_{Ar}$), 122.8 (C$_q$, C$_{Ar}$), 114.9 (+, C$_{Ar}$H), 111.7 (+, C$_{Ar}$H), 102.8 (+, C$_{Ar}$H), 55.6 (+, OCH$_3$) ppm. – **IR** (ATR, $\tilde{v}$) = 3431 (w), 3408 (m), 3373 (m), 2936 (w), 2830 (w), 1611 (w), 1574 (w), 1494 (vs), 1482 (vs), 1469 (vs), 1463 (vs), 1451 (s), 1431 (vs), 1336 (m), 1302 (m), 1266 (m), 1228 (w), 1204 (vs), 1179 (s), 1153 (vs), 1125 (s), 1109 (s), 1024 (vs), 931 (w), 844 (m), 839 (s), 827 (m), 809 (vs), 778 (vs), 727 (w), 710 (w), 640 (w), 594 (vs), 557 (w), 483 (s), 442 (vs), 422 (vs), 390 (s) cm^{-1}. – **MS** (EI, 80 °C), *m/z* (%): 227 [M]$^+$. – **HRMS** (EI, 80 °C) calc. for C$_{14}$H$_{13}$O$_2$N$_1$ [M]$^+$ 227.0946, found 227.0946. The analytical data is consistent with literature.[173]

trans-anti-9,9'-Bis(4-(4,6-bis(3,5-di-tert-butylphenyl)-1,3,5-triazin-2-yl)phenyl)-6,6'-bis(3,6-dimethoxy-9N-carbazolyl)[2.2]-1,4-1',4'-dicarbazolocyclophane (155)

Under argon atmosphere, a mixture of *trans-anti*-9,9'-bis(4-(4,6-bis(3,5-di-tert-butylphenyl)-1,3,5-triazin-2-yl)phenyl)-6,6'-di(bromo)[2.2]-1,4-1',4'-dicarbazolo-cyclophane (40.2 mg, 25.0 μmol, 1.00 equiv.), 3,6-dimethoxy-9N-carbazole (12.5 mg, 55.0 μmol, 2.20 equiv.), Pd$_2$(dba)$_3$ (2.3 mg, 2.51 μmol, 10 mol%), tri-*tert*-butylphosphonium tetrafluoroborate (1.5 mg, 5.17 μmol, 20 mol%) and sodium *tert*-butoxide (24.0 mg, 214 μmol, 8.55 equiv.) in toluene (4.0 mL) was stirred at 100 °C for 12 h. The mixture was cooled to room temperature and diluted with dichloromethane (50 mL). Then it was washed with brine (3 × 50 mL). The organic layer was dried over MgSO$_4$ and the solvent was removed under reduced pressure. The obtained crude product was purified *via* column chromatography on silica gel (cyclohexane/dichloromethane = 2:1 to 1:1) to yield the title compound as a yellow luminescent solid (35.0 mg, 18.4 μmol, 74%).

R$_f$ = 0.15 (cyclohexane/dichloromethane = 2:1). – **^{1}H NMR** (400 MHz, CDCl$_3$) δ = 9.09 (d, J = 91.7 Hz, 4H), 8.71 (d, J = 1.9 Hz, 8H), 8.33 (bs, 2H), 8.16 (d, J = 2.0 Hz, 2H), 7.78 (d, J = 8.7 Hz, 2H), 7.73 (t, J = 1.9 Hz, 4H), 7.62–7.52 (m, 8H), 7.35 (bs, 4H), 7.08 (bs, 4H), 6.21 (d, J = 7.5 Hz, 2H), 5.94 (d, J = 7.6 Hz, 2H), 3.95 (s, 12H), 3.66 (dd, J = 13.5 Hz, J = 9.3 Hz, 2H), 3.11 (dt, J = 17.2 Hz, J = 8.9 Hz, 2H), 2.96 (dd, J = 13.8 Hz, J = 9.2 Hz, 2H), 2.62 (dt, J = 13.6 Hz, J = 9.0 Hz, 2H), 1.49 (s, 72H) ppm. – **^{13}C NMR** (101 MHz, CDCl$_3$) δ = 172.4 (C$_q$, C$_{Ar}$), 170.7 (C$_q$, C$_{Ar}$), 154.0 (C$_q$, C$_{Ar}$), 151.4 (C$_q$, C$_{Ar}$), 142.2 (C$_q$, C$_{Ar}$), 141.1 (C$_q$, C$_{Ar}$), 139.8 (C$_q$, C$_{Ar}$), 137.6 (C$_q$, C$_{Ar}$), 135.7 (C$_q$, C$_{Ar}$), 135.6 (C$_q$, C$_{Ar}$), 134.7 (C$_q$, C$_{Ar}$), 131.2 (C$_q$, C$_{Ar}$), 130.5 (+, C$_{Ar}$H), 129.5 (+, C$_{Ar}$H), 127.2 (+, C$_{Ar}$H), 126.5 (C$_q$, C$_{Ar}$), 125.3 (C$_q$, C$_{Ar}$), 124.9 (+, C$_{Ar}$H), 123.4 (+, C$_{Ar}$H), 122.6 (C$_q$, C$_{Ar}$), 121.2 (+, C$_{Ar}$H), 115.4 (+, C$_{Ar}$H), 111.0 (+, C$_{Ar}$H), 110.6 (+, C$_{Ar}$H), 103.2 (+, C$_{Ar}$H), 56.4 (+, OCH$_3$), 35.2 (C$_q$, CCH$_3$), 33.6 (–, CH$_2$), 31.6 (+, CH$_3$), 30.3 (–, CH$_2$) ppm. – **IR** (ATR, ṽ) = 2952 (w), 1601 (w), 1579 (w), 1516 (vs), 1489 (s), 1475 (vs), 1434 (s), 1366 (s), 1357 (s), 1283 (w), 1248 (w), 1205 (vs), 1153 (s), 1037 (w), 895 (w), 829 (s), 820 (w), 701 (w), 683 (w), 626 (w), 578 (w), 551 (w), 493 (w) cm^{-1}. – **MS** (FAB, 3-NBA), *m/z* (%): 1900 [M]$^+$.

5.5 Crystal Structures

5.5.1 Crystallographic Data Solved by Dr. Martin Nieger

Crystal structures in this section were measured and solved by Dr. Martin Nieger at the University of Helsinki.

Table 33. Overview over the numbering and sample coding of crystals from Dr. Nieger.

Numbering in this thesis	Sample Code used by Dr. Nieger
53	SB1313_hy
57	SB1316_hy
45	SB1044_hy
46	SB1295_hy
49	SB1097_hy
95	SB1290_hy
125	SB1273_hy
152	SB1266_hy
133	ZZ-552

10-(3,5-Dichlorophenyl)-10*N*-phenoxazine (53) – SB1313_hy

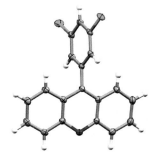

$C_{18}H_{11}Cl_2NO$	$F(000) = 672$
$M_r = 328.18$	$D_x = 1.495$ Mg m^{-3}
Monoclinic, $P2_1/n$ (no.14)	Cu $K\alpha$ radiation, $\lambda = 1.54178$ Å
$a = 9.3067$ (6) Å	Cell parameters from 9922 reflections
$b = 8.4136$ (5) Å	$\theta = 5.4–72.1°$
$c = 18.6751$ (12) Å	$\mu = 4.00$ mm^{-1}
$\beta = 94.587$ (2)°	$T = 123$ K
$V = 1457.63$ (16) Å^3	Plates, colourless
$Z = 4$	$0.20 \times 0.12 \times 0.04$ mm

Data collection

Bruker D8 VENTURE diffractometer with PhotonII CPAD detector	2810 reflections with $I > 2\sigma(I)$
Radiation source: INCOATEC microfocus sealed tube	$R_{int} = 0.026$
rotation in ϕ and ω, 1°, shutterless scans	$\theta_{max} = 72.2°$, $\theta_{min} = 4.8°$
Absorption correction: multi-scan SADABS (Sheldrick, 2014)	$h = -11 \rightarrow 9$
$T_{min} = 0.665$, $T_{max} = 0.841$	$k = -10 \rightarrow 10$
15073 measured reflections	$l = -22 \rightarrow 23$
2877 independent reflections	

Refinement

Refinement on F^2	Primary atom site location: dual
Least-squares matrix: full	Secondary atom site location: difference Fourier map
$R[F^2 > 2\sigma(F^2)] = 0.026$	Hydrogen site location: difference Fourier map
$wR(F^2) = 0.070$	H-atom parameters constrained
$S = 1.07$	$w = 1/[\sigma^2(F_o^2) + (0.0347P)^2 + 0.5822P]$ where $P = (F_o^2 + 2F_c^2)/3$
2877 reflections	$(\Delta/\sigma)_{max} < 0.001$
199 parameters	$\Delta\rangle_{max} = 0.24$ e Å^{-3}
0 restraints	$\Delta\rangle_{min} = -0.25$ e Å^{-3}

10-(3-Chloro-4-methoxyphenyl)-10N-phenoxazine (57) – SB1316_hy

$C_{19}H_{14}ClNO_2$	$F(000) = 672$
$M_r = 323.76$	$D_x = 1.427$ Mg m^{-3}
Monoclinic, $P2_1/n$ (no.14)	Cu $K\alpha$ radiation, $\lambda = 1.54178$ Å
$a = 8.5998$ (3) Å	Cell parameters from 9976 reflections
$b = 8.0062$ (3) Å	$\theta = 4.0$–$72.3°$
$c = 21.9748$ (9) Å	$\mu = 2.32$ mm^{-1}
$\beta = 95.064$ (1)°	$T = 123$ K
$V = 1507.10$ (10) Å^3	Blocks, colourless
$Z = 4$	$0.20 \times 0.16 \times 0.10$ mm

Data collection

Bruker D8 VENTURE diffractometer with PhotonII CPAD detector	2908 reflections with $I > 2\sigma(I)$
Radiation source: INCOATEC microfocus sealed tube	$R_{int} = 0.024$
rotation in ϕ and ω, 1°, shutterless scans	$\theta_{max} = 72.3°$, $\theta_{min} = 4.0°$
Absorption correction: multi-scan $SADABS$ (Sheldrick, 2014)	$h = -9 \rightarrow 10$
$T_{min} = 0.696$, $T_{max} = 0.795$	$k = -9 \rightarrow 9$
15246 measured reflections	$l = -23 \rightarrow 27$
2969 independent reflections	

Refinement

Refinement on F^2	Primary atom site location: dual
Least-squares matrix: full	Secondary atom site location: difference Fourier map
$R[F^2 > 2\sigma(F^2)] = 0.028$	Hydrogen site location: difference Fourier map
$wR(F^2) = 0.077$	H-atom parameters constrained
$S = 1.05$	$w = 1/[\sigma^2(F_o^2) + (0.0397P)^2 + 0.6138P]$ where $P = (F_o^2 + 2F_c^2)/3$
2969 reflections	$(\Delta/\sigma)_{max} = 0.001$
209 parameters	$\Delta\rangle_{max} = 0.31$ e Å^{-3}
0 restraints	$\Delta\rangle_{min} = -0.26$ e Å^{-3}

10-(8,9-Dioxa-8a-borabenzo[fg]tetracen-2-yl)-10N-phenoxazine (45) – SB1044_hy

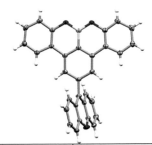

$C_{30}H_{18}BNO_3$	$Z = 2$
$M_r = 451.26$	$F(000) = 468$
Triclinic, $P\text{-}1$ (no.2)	$D_x = 1.437$ Mg m^{-3}
$a = 5.9979$ (3) Å	Cu $K\alpha$ radiation, $\lambda = 1.54178$ Å
$b = 13.5277$ (6) Å	Cell parameters from 9852 reflections
$c = 14.3850$ (7) Å	$\theta = 3.4$–$72.2°$
$\alpha = 64.594$ (2)°	$\mu = 0.74$ mm^{-1}
$\beta = 81.973$ (2)°	$T = 123$ K
$\gamma = 84.275$ (2)°	Plates, yellow
$V = 1042.93$ (9) Å^3	$0.24 \times 0.16 \times 0.04$ mm

Data collection

Bruker D8 VENTURE diffractometer with Photon100 detector	4107 independent reflections
Radiation source: INCOATEC microfocus sealed tube	3686 reflections with $I > 2\sigma(I)$
Detector resolution: 10.4167 pixels mm^{-1}	$R_{int} = 0.025$
rotation in ϕ and ω, 1°, shutterless scans	$\theta_{max} = 72.2°$, $\theta_{min} = 3.4°$
Absorption correction: multi-scan *SADABS* (Sheldrick, 2014)	$h = -7{\rightarrow}7$
$T_{min} = 0.875$, $T_{max} = 0.971$	$k = -16{\rightarrow}16$
18449 measured reflections	$l = -17{\rightarrow}17$

Refinement

Refinement on F^2	Secondary atom site location: difference Fourier map
Least-squares matrix: full	Hydrogen site location: difference Fourier map
$R[F^2 > 2\sigma(F^2)] = 0.039$	H-atom parameters constrained
$wR(F^2) = 0.108$	$w = 1/[\sigma^2(F_o^2) + (0.0528P)^2 + 0.4041P]$ where $P = (F_o^2 + 2F_c^2)/3$
$S = 1.06$	$(\Delta/\sigma)_{max} = 0.001$
4107 reflections	$\Delta\rangle_{max} = 0.73$ e Å^{-3}
317 parameters	$\Delta\rangle_{min} = -0.19$ e Å^{-3}
0 restraints	Extinction correction: *SHELXL2014*/7 (Sheldrick 2014, Fc'=kFc[1+0.001xFc²λ³/sin(2θ)]$^{-1/4}$
Primary atom site location: dual	Extinction coefficient: 0.0022 (4)

10-(8,9-Dioxa-8a-borabenzo[fg]tetracen-12-yl)-10H-phenoxazine (46) – SB1295_hy

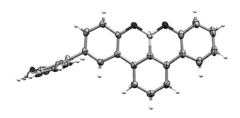

$C_{30}H_{18}BNO_3$	$D_x = 1.400$ Mg m^{-3}
$M_r = 451.26$	Cu $K\alpha$ radiation, $\lambda = 1.54178$ Å
Orthorhombic, $Pbca$ (no.61)	Cell parameters from 9925 reflections
$a = 9.0398$ (2) Å	$\theta = 5.1-72.0°$
$b = 17.2238$ (4) Å	$\mu = 0.72$ mm^{-1}
$c = 27.4998$ (6) Å	$T = 123$ K
$V = 4281.71$ (17) Å^3	Plates, yellow
$Z = 8$	$0.16 \times 0.08 \times 0.03$ mm
$F(000) = 1872$	

Data collection

Bruker D8 VENTURE diffractometer with PhotonII CPAD detector	3625 reflections with $I > 2\sigma(I)$
Radiation source: INCOATEC microfocus sealed tube	$R_{int} = 0.048$
rotation in ϕ and ω, 1°, shutterless scans	$\theta_{max} = 72.1°$, $\theta_{min} = 3.2°$
Absorption correction: multi-scan $SADABS$ (Sheldrick, 2014)	$h = -9 \rightarrow 11$
$T_{min} = 0.843$, $T_{max} = 0.971$	$k = -18 \rightarrow 21$
39582 measured reflections	$l = -33 \rightarrow 33$
4205 independent reflections	

Refinement

Refinement on F^2	Primary atom site location: dual
Least-squares matrix: full	Secondary atom site location: difference Fourier map
$R[F^2 > 2\sigma(F^2)] = 0.048$	Hydrogen site location: difference Fourier map
$wR(F^2) = 0.129$	H-atom parameters constrained
$S = 1.05$	$w = 1/[\sigma^2(F_o^2) + (0.0604P)^2 + 3.0872P]$ where $P = (F_o^2 + 2F_c^2)/3$
4205 reflections	$(\Delta/\sigma)_{max} < 0.001$
316 parameters	$\Delta\rangle_{max} = 0.84$ e Å^{-3}
0 restraints	$\Delta\rangle_{min} = -0.19$ e Å^{-3}

10-(8,9-Dioxa-8a-borabenzo[fg]tetracen-12-yl)-9,9-dimethyl-9,10-dihydroacridine (49) – SB1097_hy

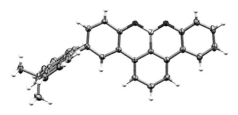

$C_{33}H_{24}BNO_2$	$Z = 4$
$M_r = 477.34$	$F(000) = 1000$
Triclinic, P-1 (no.2)	$D_x = 1.333$ Mg m^{-3}
$a = 8.8265$ (3) Å	Cu $K\alpha$ radiation, $\lambda = 1.54178$ Å
$b = 13.4334$ (4) Å	Cell parameters from 9826 reflections
$c = 20.8908$ (6) Å	$\theta = 3.3$–$72.2°$
$\alpha = 76.063$ (1)°	$\mu = 0.64$ mm^{-1}
$\beta = 83.015$ (1)°	$T = 123$ K
$\gamma = 84.063$ (1)°	Blocks, colourless
$V = 2379.18$ (13) Å^3	$0.24 \times 0.12 \times 0.06$ mm

Data collection

Bruker D8 VENTURE diffractometer with PhotonII CPAD detector	8583 reflections with $I > 2\sigma(I)$
Radiation source: INCOATEC microfocus sealed tube	$R_{int} = 0.024$
rotation in ϕ and ω, 1°, shutterless scans	$\theta_{max} = 72.3°$, $\theta_{min} = 3.4°$
Absorption correction: multi-scan $SADABS$ (Sheldrick, 2014)	$h = -10 \rightarrow 10$
$T_{min} = 0.853$, $T_{max} = 0.971$	$k = -16 \rightarrow 16$
34843 measured reflections	$l = -25 \rightarrow 25$
9321 independent reflections	

Refinement

Refinement on F^2	Primary atom site location: dual
Least-squares matrix: full	Secondary atom site location: difference Fourier map
$R[F^2 > 2\sigma(F^2)] = 0.037$	Hydrogen site location: difference Fourier map
$wR(F^2) = 0.102$	H-atom parameters constrained
$S = 1.03$	$w = 1/[\sigma^2(F_o^2) + (0.0507P)^2 + 0.7154P]$ where $P = (F_o^2 + 2F_c^2)/3$
9321 reflections	$(\Delta/\sigma)_{max} = 0.001$
667 parameters	$\Delta\rangle_{max} = 0.31$ e Å^{-3}
0 restraints	$\Delta\rangle_{min} = -0.19$ e Å^{-3}

2,4-Bis(4-(*tert*-butyl)phenyl)-6-(3,4,5-trifluorophenyl)-1,3,5-triazine (95) – SB1290_hy

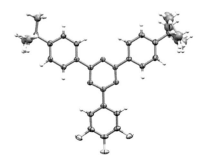

$C_{29}H_{28}F_3N_3$	$Z = 2$
$M_r = 475.54$	$F(000) = 500$
Triclinic, P-1 (no.2)	$D_x = 1.253$ Mg m^{-3}
$a = 6.6136$ (2) Å	Cu $K\alpha$ radiation, $\lambda = 1.54178$ Å
$b = 12.2094$ (4) Å	Cell parameters from 7031 reflections
$c = 16.2642$ (6) Å	$\theta = 2.7–72.1°$
$\alpha = 94.760$ (2)°	$\mu = 0.74$ mm^{-1}
$\beta = 94.366$ (2)°	$T = 123$ K
$\gamma = 104.550$ (2)°	Plates, colourless
$V = 1260.38$ (7) Å^3	$0.22 \times 0.08 \times 0.02$ mm

Data collection

Bruker D8 VENTURE diffractometer with PhotonII CPAD detector	4005 reflections with $I > 2\sigma(I)$
Radiation source: INCOATEC microfocus sealed tube	$R_{int} = 0.033$
rotation in ϕ and ω, 1°, shutterless scans	$\theta_{max} = 72.1°$, $\theta_{min} = 2.7°$
Absorption correction: multi-scan *SADABS* (Sheldrick, 2014)	$h = -8\rightarrow8$
$T_{min} = 0.778$, $T_{max} = 0.971$	$k = -15\rightarrow14$
14766 measured reflections	$l = -19\rightarrow20$
4899 independent reflections	

Refinement

Refinement on F^2	Primary atom site location: dual
Least-squares matrix: full	Secondary atom site location: difference Fourier map
$R[F^2 > 2\sigma(F^2)] = 0.072$	Hydrogen site location: mixed
$wR(F^2) = 0.198$	H-atom parameters constrained
$S = 1.05$	$w = 1/[\sigma^2(F_o^2) + (0.0821P)^2 + 1.5111P]$ where $P = (F_o^2 + 2F_c^2)/3$
4899 reflections	$(\Delta/\sigma)_{max} < 0.001$
306 parameters	$\Delta\rangle_{max} = 0.75$ e Å^{-3}
90 restraints	$\Delta\rangle_{min} = -0.58$ e Å^{-3}

5,11-Bis(4-chloro-2,6-dimethylphenyl)-5,11-dihydroindolo[3,2-b]carbazole(125) – SB1273_hy

$C_{34}H_{26}Cl_2N_2$	$F(000) = 556$
$M_r = 533.47$	$D_x = 1.330$ Mg m^{-3}
Monoclinic, $P2_1/c$ (no.14)	Mo $K\alpha$ radiation, $\lambda = 0.71073$ Å
$a = 11.9854$ (5) Å	Cell parameters from 8313 reflections
$b = 7.9711$ (3) Å	$\theta = 2.8–27.5°$
$c = 14.8085$ (6) Å	$\mu = 0.27$ mm^{-1}
$\beta = 109.643$ (2)°	$T = 123$ K
$V = 1332.43$ (9) Å^3	Plates, yellow
$Z = 2$	$0.32 \times 0.10 \times 0.04$ mm

Data collection

Bruker D8 VENTURE diffractometer with PhotonII CPAD detector	2613 reflections with $I > 2\sigma(I)$
Radiation source: INCOATEC microfocus sealed tube	$R_{int} = 0.038$
rotation in ϕ, 1°, shutterless scans	$\theta_{max} = 27.6°, \theta_{min} = 2.9°$
Absorption correction: multi-scan *SADABS* (Sheldrick, 2014)	$h = -15 \rightarrow 15$
$T_{min} = 0.925, T_{max} = 0.977$	$k = -10 \rightarrow 10$
19441 measured reflections	$l = -19 \rightarrow 19$
3068 independent reflections	

Refinement

Refinement on F^2	Primary atom site location: dual
Least-squares matrix: full	Secondary atom site location: difference Fourier map
$R[F^2 > 2\sigma(F^2)] = 0.042$	Hydrogen site location: difference Fourier map
$wR(F^2) = 0.116$	H-atom parameters constrained
$S = 1.05$	$w = 1/[\sigma^2(F_o^2) + (0.0508P)^2 + 0.9772P]$ where $P = (F_o^2 + 2F_c^2)/3$
3068 reflections	$(\Delta/\sigma)_{max} < 0.001$
174 parameters	$\Delta\rangle_{max} = 1.10$ e Å^{-3}
0 restraints	$\Delta\rangle_{min} = -0.31$ e Å^{-3}

4, 16-*N*-Di(2-chlorophenyl)amino[2.2]paracyclophane (152) – SB1266_hy

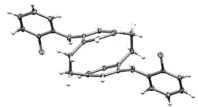

$C_{28}H_{24}Cl_2N_2$	$D_x = 1.421$ Mg m^{-3}
$M_r = 459.39$	Cu $K\alpha$ radiation, $\lambda = 1.54178$ Å
Orthorhombic, *Pbca (no.61)*	Cell parameters from 9772 reflections
$a = 12.3953$ (7) Å	$\theta = 5.0$–72.2°
$b = 7.0720$ (4) Å	$\mu = 2.86$ mm^{-1}
$c = 24.4985$ (13) Å	$T = 123$ K
$V = 2147.5$ (2) Å^3	Plates, yellow
$Z = 4$	$0.12 \times 0.06 \times 0.02$ mm
$F(000) = 960$	

Data collection

Bruker D8 VENTURE diffractometer with PhotonII CPAD detector	2039 reflections with $I > 2\sigma(I)$
Radiation source: INCOATEC microfocus sealed tube	$R_{int} = 0.028$
rotation in ϕ and ω, 1°, shutterless scans	$\theta_{max} = 72.2°, \theta_{min} = 3.6°$
Absorption correction: multi-scan *SADABS* (Sheldrick, 2014)	$h = -15 \rightarrow 15$
$T_{min} = 0.788, T_{max} = 0.915$	$k = -7 \rightarrow 8$
27592 measured reflections	$l = -30 \rightarrow 28$
2126 independent reflections	

Refinement

Refinement on F^2	Primary atom site location: dual
Least-squares matrix: full	Secondary atom site location: difference Fourier map
$R[F^2 > 2\sigma(F^2)] = 0.029$	Hydrogen site location: difference Fourier map
$wR(F^2) = 0.075$	H atoms treated by a mixture of independent and constrained refinement
$S = 1.05$	$w = 1/[\sigma^2(F_o^2) + (0.0355P)^2 + 1.4614P]$ where $P = (F_o^2 + 2F_c^2)/3$
2126 reflections	$(\Delta/\sigma)_{max} = 0.002$
148 parameters	$\Delta\rangle_{max} = 0.28$ e Å^{-3}
1 restraint	$\Delta\rangle_{min} = -0.24$ e Å^{-3}

(*rac*)-1-(*N*-[2]Paracyclo[2]-6-(rifluoromethyl)(1,4)carbazolophanyl)-4-(4,6-diphenyl-1,3,5-triazin-2-yl)-benzene (133)_ZZ-552

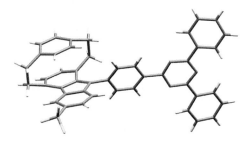

C₄₄H₃₁F₃N₄	$Z = 8$
$M_r = 672.73$	$F(000) = 2800$
Monoclinic, C2/c (no.15)	$D_x = 1.351$ Mg m⁻³
$a = 21.7598$ (7) Å	Cu $K\alpha$ radiation, $\lambda = 1.54178$ Å
$b = 9.5073$ (3) Å	$\mu = 0.75$ mm⁻¹
$c = 31.9788$ (8) Å	$T = 123$ K
$\beta = 90.574$ (2)°	0.20 × 0.14 × 0.08 mm
$V = 6615.3$ (3) Å³	

Due to the bad quality of the crystal, only the constitution and conformation could be determined.

5.5.2 Crystallographic Data Solved by Dr. Olaf Fuhr

Crystal structures in this section were measured and solved by Dr. Olaf Fuhr at the Institute of Nanotechnology (INT) at the Karlsruhe Institute of Technology.

Table 34. Overview over the numbering and sample coding of crystals from Dr. Fuhr.

Numbering in this thesis	Sample Code used by Dr. Fuhr
109	ZZ-229
110	ZZ-264
111	ZZ-267
112	ZZ-280
148	ZZ-309
156	ZZ-469

5,11-Bis(2,6-dimethyl-4-(5-methylpyridin-2-yl)phenyl)-5,11-dihydroindolo[3,2-b]carbazole (109) – ZZ-229

Crystal data and structure refinement for Z229.

Identification code	Z229
Empirical formula	$C_{46}H_{38}N_4$
Formula weight	646.80
Temperature/K	180.15
Crystal system	monoclinic
Space group	C2/c
a/Å	29.1702(10)
b/Å	8.3113(2)
c/Å	16.6047(6)
α/°	90
β/°	120.274(2)
γ/°	90
Volume/Å³	3476.7(2)
Z	4
ρ_{calc}g/cm³	1.236
μ/mm⁻¹	0.359
F(000)	1368.0
Crystal size/mm³	0.19 × 0.16 × 0.03
Radiation	GaKα (λ = 1.34143)
2Θ range for data collection/°	6.104 to 128.122
Index ranges	-38 ≤ h ≤ 38, -10 ≤ k ≤ 3, -21 ≤ l ≤ 21
Reflections collected	17172
Independent reflections	4234 [R_{int} = 0.0366, R_{sigma} = 0.0270]
Indep. refl. with I>=2σ (I)	2913
Data/restraints/parameters	4234/0/229
Goodness-of-fit on F^2	1.066
Final R indexes [I>=2σ (I)]	R_1 = 0.0547, wR_2 = 0.1602
Final R indexes [all data]	R_1 = 0.0772, wR_2 = 0.1721
Largest diff. peak/hole / e Å⁻³	0.36/-0.38

5,11-Bis(2,6-dimethyl-4-(5-methylpyrimidin-2-yl)phenyl)-5,11-dihydroindolo[3,2-b]carbazole (110) – ZZ-264

Crystal data and structure refinement for ZZ-264.

Identification code	ZZ-264
Empirical formula	$C_{46}H_{38}Cl_6N_6$
Formula weight	887.52
Temperature/K	180.15
Crystal system	triclinic
Space group	P-1
a/Å	9.4988(4)
b/Å	10.4665(4)
c/Å	11.8330(4)
α/°	90.956(3)
β/°	112.343(3)
γ/°	99.639(3)
Volume/Å^3	1068.57(7)
Z	1
ρ_{calc}g/cm^3	1.379
μ/mm^{-1}	2.607
F(000)	458.0
Crystal size/mm^3	0.24 × 0.23 × 0.2
Radiation	GaKα (λ = 1.34143)
2Θ range for data collection/°	7.054 to 125.098
Index ranges	-6 ≤ h ≤ 12, -13 ≤ k ≤ 13, -15 ≤ l ≤ 15
Reflections collected	12789
Independent reflections	4950 [R_{int} = 0.0160, R_{sigma} = 0.0149]
Indep. refl. with I>=2σ (I)	4045
Data/restraints/parameters	4950/0/265
Goodness-of-fit on F^2	1.080
Final R indexes [I>=2σ (I)]	R_1 = 0.0516, wR_2 = 0.1452
Final R indexes [all data]	R_1 = 0.0604, wR_2 = 0.1512
Largest diff. peak/hole / e Å^{-3}	1.44/-0.47

5,11-Bis(2,6-dimethyl-4-(5-(trifluoromethyl)pyrimidin-2-yl)phenyl)-5,11-dihydroindolo[3,2-b]carbazole (111) – ZZ-267

Crystal data and structure refinement for ZZ-267.

Identification code	ZZ-267
Empirical formula	$C_{44}H_{30}F_6N_6$
Formula weight	756.74
Temperature/K	180.15
Crystal system	monoclinic
Space group	$P2_1/n$
a/Å	8.0923(7)
b/Å	38.238(4)
c/Å	11.7771(8)
α/°	90
β/°	96.543(6)
γ/°	90
Volume/Å³	3620.5(5)
Z	4
ρ_{calc}g/cm³	1.388
μ/mm⁻¹	0.567
F(000)	1560.0
Crystal size/mm³	0.26 × 0.23 × 0.05
Radiation	GaKα (λ = 1.34143)
2Θ range for data collection/°	4.02 to 124.978
Index ranges	-9 ≤ h ≤ 10, -50 ≤ k ≤ 36, -15 ≤ l ≤ 7
Reflections collected	23946
Independent reflections	8411 [R_{int} = 0.0611, R_{sigma} = 0.0796]
Indep. refl. with I>=2σ (I)	3824
Data/restraints/parameters	8411/0/510
Goodness-of-fit on F²	0.878
Final R indexes [I>=2σ (I)]	R_1 = 0.0578, wR_2 = 0.1451
Final R indexes [all data]	R_1 = 0.1251, wR_2 = 0.1637
Largest diff. peak/hole / e Å⁻³	0.27/-0.30

5,11-Bis(2,6-dimethyl-4-(5-(trifluoromethyl)pyrazin-2-yl)phenyl)-5,11-dihydroindolo[3,2-b]carbazole (112) – ZZ-280

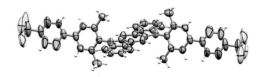

Crystal data and structure refinement for ZZ-280.

Identification code	ZZ-280
Empirical formula	$C_{44}H_{30}F_6N_6$
Formula weight	756.74
Temperature/K	180.15
Crystal system	triclinic
Space group	P-1
a/Å	8.2271(5)
b/Å	11.6078(7)
c/Å	19.1533(12)
α/°	92.690(5)
β/°	102.630(5)
γ/°	95.831(5)
Volume/Å³	1771.11(19)
Z	2
ρ_{calc}g/cm³	1.419
μ/mm⁻¹	0.579
F(000)	780.0
Crystal size/mm³	0.32 × 0.29 × 0.03
Radiation	GaKα (λ = 1.34143)
2Θ range for data collection/°	6.676 to 109.992
Index ranges	-10 ≤ h ≤ 10, -5 ≤ k ≤ 14, -23 ≤ l ≤ 23
Reflections collected	17399
Independent reflections	6566 [R_{int} = 0.0845, R_{sigma} = 0.0698]
Indep. refl. with I>=2σ (I)	2882
Data/restraints/parameters	6566/1/509
Goodness-of-fit on F^2	1.419
Final R indexes [I>=2σ (I)]	R_1 = 0.1366, wR_2 = 0.3183
Final R indexes [all data]	R_1 = 0.2170, wR_2 = 0.3475
Largest diff. peak/hole / e Å⁻³	0.66/-0.40

trans-anti-[2.2]-1,4-1',4'-dicarbazolocyclophane (148, dicarbazolocyclophane, DCCP) – ZZ-309

Crystal data and structure refinement for ZZ-309.

Identification code	ZZ-309
Empirical formula	$C_{32}H_{34}N_2O_2S_2$
Formula weight	542.73
Temperature/K	180.15
Crystal system	orthorhombic
Space group	Pbca
a/Å	9.3241(3)
b/Å	16.4494(5)
c/Å	17.6438(5)
α/°	90
β/°	90
γ/°	90
Volume/Å^3	2706.13(14)
Z	4
ρ_{calc}g/cm^3	1.332
μ/mm^{-1}	1.340
F(000)	1152.0
Crystal size/mm^3	0.21 × 0.2 × 0.19
Radiation	GaKα (λ = 1.34143)
2Θ range for data collection/°	9.356 to 125.002
Index ranges	-12 ≤ h ≤ 11, -21 ≤ k ≤ 12, -21 ≤ l ≤ 23
Reflections collected	15304
Independent reflections	3236 [R_{int} = 0.0252, R_{sigma} = 0.0198]
Indep. refl. with I>=2σ (I)	2650
Data/restraints/parameters	3236/0/240
Goodness-of-fit on F^2	1.104
Final R indexes [I>=2σ (I)]	R_1 = 0.0409, wR_2 = 0.1086
Final R indexes [all data]	R_1 = 0.0496, wR_2 = 0.1126
Largest diff. peak/hole / e Å^{-3}	0.29/-0.35

trans-anti-6,6'-Dibromo[2.2]-1,4-1',4'-dicarbazolocyclophane (156) – ZZ-469

Crystal data and structure refinement for ZZ469.

Identification code	ZZ469
Empirical formula	$C_{36}H_{36}Br_2N_2O_2$
Formula weight	688.49
Temperature/K	150
Crystal system	monoclinic
Space group	$P2_1/n$
a/Å	9.9198(3)
b/Å	15.7283(4)
c/Å	10.3657(3)
α/°	90
β/°	112.308(2)
γ/°	90
Volume/Å^3	1496.23(8)
Z	2
$\rho_{calc}g/cm^3$	1.528
μ/mm^{-1}	2.552
F(000)	704.0
Crystal size/mm^3	0.18 × 0.16 × 0.14
Radiation	GaKα (λ = 1.34143)
2Θ range for data collection/°	9.142 to 124.976
Index ranges	-11 ≤ h ≤ 13, -20 ≤ k ≤ 20, -13 ≤ l ≤ 5
Reflections collected	15306
Independent reflections	3575 [R_{int} = 0.0336, R_{sigma} = 0.0229]
Data/restraints/parameters	3575/0/190
Goodness-of-fit on F^2	1.097
Final R indexes [I>=2σ (I)]	R_1 = 0.0556, wR_2 = 0.1570
Final R indexes [all data]	R_1 = 0.0608, wR_2 = 0.1647
Largest diff. peak/hole / e Å^{-3}	1.54/-0.97

6 List of Abbreviations

°C	degree celsius
ΔE_{ST}	energy gap between S_1 and T_1
δ	chemical shift
μmol	micromole
Å	Ångström
A	Acceptor
Alq$_3$	tris(8-hydroxyquinolinato)aluminium
APCI	atmospheric-pressure chemical ionization
a.q.	aqueous
Ar	aromat(ic)
ATR	attenuated total reflection
ATRP	atom transfer radical polymerization
a.u.	arbitrary unit
Bu	butyl
Bz	Benzoyl
c.w.	clockwise
calc.	calculated
CBP	4,4'-bis-(N-carbazolyl)-1,1'-biphenyl
CE$_{max}$	maximal current efficiency
CIE	*commission internationale de l'éclairage*
CPL	circularly polarized luminescence
CT	charge transfer
Cz	carbazole
Czp	[2]Paracyclo[2](1,4)carbazolophane
d	doublet
D	Donor

dba	dibenzylideneacetone
DCM	dichloromethane
DCCP	*trans-anti*-[2.2]-1,4-1',4'-dicarbazolocyclophane
DEPT	distortionless enhancement by polarization transfer
DF	delayed fluorescence
DFT	density functional theory
DMAC	dimethylacridan
DMF	*N,N*-dimethylformamide
DMSO	dimethyl sulfoxide
DPEPO	bis[2-(diphenylphosphino)phenyl] ether oxide
DSC	differential scanning calorimetry
E	electrophile
e.g.	exempli gratia (for example)
EA	elemental analysis
EBL	electron blocking layer
EDG	electron donating group
EI	electron ionization
EIL	electon injection layer
EL	electroluminescence
EML	emissive layer
EQE	external quantum efficiency
equiv.	equivalents
Et	ethyl
et al.	*et alii* (and others)
etc.	*et cetera* (and so on)
ETL	electron transport layer
EWG	electron withdrawing group
eV	electron volt

f	oscillator strength
FAB	fast atom bombardment
g	gram
h	hour
h	Planck constant
HBL	hole blocking layer
HIL	hole injection layer
HOMO	highest occupied molecular orbital
HRMS	high resolution mass spectrometry
HTL	hole transport layer
Hz	Hertz
ICT	intramolecular charge transfer
ICz	5,11-dihydroindolo[3,2-b]carbazole
IR	infrared
ISC	intersystem crossing
ITO	indium tin oxide
J	coupling constant
K	Kelvin
L	liter
LC	liquid crystal
LED	light-emitting diode
lm	lumen
LUMO	lowest unoccupied molecular orbital
m	*meta*
M	molar/liter
m	multiplet
mbar	millibar
mCP	1,3-bis-(N-carbazolyl)benzene

Me	methyl
mg	milligram
MHz	mega Hertz
min	minute
mL	milliliter
mol	millimole
n-BuLi	n-butyllithium
NBS	*N*-bromosuccinimide
NPB	*N,N'*-bis(naphthalen-1-yl)-*N,N'*-bis(phenyl)benzidine
NMR	nuclear magnetic resonance
MS	mass spectrometry
Nu	Nucleophile
OBO	OBO-fused benzo[fg]tetracene
OLED	organic light-emitting diode
o	*ortho*
p	*para*
PCP	[2.2]paracyclophane
PE_{max}	maximum power efficiency
PEDOT:PSS	poly(3,4-ethylenedioxythiophene) polystyrene sulfonate
PF	prompt fluorescence
Ph	phenyl
Piv	pivalic
PL	photoluminescence
PLQY	photoluminescence quantum yield
PMMA	poly(methyl methacrylate)
ppm	parts per million
PPT	2,8-bis(diphenylphosphoryl)dibenzo[b,d]thiophene
PXZ	phenoxazine

Py	pyridyl
q	quartet
R/R_p	right-handed (clockwise) stereodescriptor
r.t.	room temperature
rac	racemic
RISC	reverse intersystem crossing
s	seconds
s	singlet
S_1	first excitet singlet state
S/S_p	left-handed (counter-clockwise) stereodescriptor
sat.	saturated
SCE	saturated calomel electrode
SEM	Scanning Electron Microscopy
SPhos	2-dicyclohexylphosphino-2',6'-dimethoxybiphenyl
t	triplet
T	temperature
T_1	first excited triplet state
TADF	thermally activated delayed fluorescence
^{t}Bu	*tert*-butyl
THF	tetrahydrofuran
TLC	thin layer chromatography
TmPyPB	1,3,5-tris(3-pyridyl-3-phenyl)benzene
tol	toluyl
Trz	1,3,5-triazine
UV	ultraviolet
V	volt
Vis	visible light
V_{on}	onset voltage

vs	very strong
vs	*versus*
vw	very weak
w	weak
W	watt
wt%	weight percent

7 Bibliography

[1]   J. I. Seeman, *Angew. Chem. Int. Ed.* **2016**, *55*, 12898–12912.

[2]   C. W. Tang, S. A. VanSlyke, *Appl. Phys. Lett.* **1987**, *51*, 913–915.

[3]   B. Geffroy, P. le Roy, C. Prat, *Polym. Inter.* **2006**, *55*, 572–582.

[4]   Mitsuhiro Koden, *OLED Display and Lighting*, John Wiley & Sons, **2017**.

[5]   X. H. Zhu, J. Peng, Y. Cao, J. Roncali, *Chem. Soc. Rev.* **2011**, *40*, 3509–3524.

[6]   S. Y. Lee, T. Yasuda, Y. S. Yang, Q. Zhang, C. Adachi, *Angew. Chem. Int. Ed.* **2014**, *53*, 6402–6406.

[7]   C.-J. Chiang, C. Winscom, S. Bull, A. Monkman, *Org. Electron.* **2009**, *10*, 1268–1274.

[8]   G. Gu, V. Bulović, P. E. Burrows, S. R. Forrest, M. E. Thompson, *Appl. Phys. Lett.* **1996**, *68*, 2606–2608.

[9]   Huawei OLED Phones. https://consumer.huawei.com/cn/phones/mate30-pro-5g/, **2019**.

[10]   "ISE 2018: LG puts focus on Transparent OLED," can be found under https://www.installation-international.com/technology/ise-2018-lg-puts-focus-transparent-oled, **2018**.

[11]   OLED Displays. https://www.cynora.com/technology/oled/, **2019**.

[12]   H. J. Round, *Electr. World* **1907**, 19, 309.

[13]   A. Bernanose, *Br. J. Appl. Phys.* **1955**, 6, 54–56.

[14]   M. Pope, H. P. Kallmann, P. Magnante, *J. Chem. Phys.* **1963**, *38*, 2042–2043.

[15]   F. Hundemer. *Modular Design Strategies for TADF Emitters: Towards Highly Efficient Materials for OLED Application*, Ph.D. Thesis, *Karlsruhe Institute of Technology*, **2019**.

[16]   H. Yersin, *Highly Efficient OLEDs: Materials Based on Thermally Activated Delayed Fluorescence, John Wiley & Sons,* **2018**.

[17]   C. Bizzarri, E. Spuling, D. M. Knoll, D. Volz, S. Bräse, *Coor. Chem. Rev.* **2018**, *373*, 49–82.

[18]   E. Spuling, *Synthesis of New [2.2]Paracyclophane Derivatives for Application in Material Sciences, Logos-Verlag.* **2019**.

[19]   *R.M. Metzger, Unimolecular and Supramolecular Electronics I: Chemistry and Physics Meet at Metal-Molecule Interfaces, Springer,* **2012**.

[20]   H. H. Yu, S.-J. Hwang, K.-C. Hwang, *Optical Commun.* **2005**, *248*, 51–57.

[21]   T. P. Nguyen, P. Le Rendu, N. N. Dinh, M. Fourmigué, C. Mézière, *Synth. Met.* **2003**, *138*, 229–232.

[22]   A. Benor, S.-Y. Takizawa, C. Pérez-Bolívar, P. Anzenbacher, *Org. Electron.* **2010**, *11*, 938–945.

[23]   C. Giebeler, H. Antoniadis, D. D. C. Bradley, Y. Shirota, *J. Appl. Phys.* **1999**, *85*, 608–615.

[24]   H. Hirayama, Y. Tsukada, T. Maeda, N. Kamata, *Appl. Phys. Express* **2010**, *3*, 031002.

[25] N. C. Erickson, R. J. Holmes, *Adv. Funct. Mater.* **2013**, *23*, 5190–5198.

[26] J. Luo, Z. Xie, J. W. Lam, L. Cheng, H. Chen, C. Qiu, H. S. Kwok, X. Zhan, Y. Liu, D. Zhu, B. Z. Tang, *Chem. Commun.* **2001**, 1740–1741.

[27] J. Huang, H. Nie, J. Zeng, Z. Zhuang, S. Gan, Y. Cai, J. Guo, S.-J. Su, Z. Zhao, B. Z. Tang, *Angew. Chem. Int. Ed.* **2017**, *56*, 12971–12976.

[28] H. Liu, J. Zeng, J. Guo, H. Nie, Z. Zhao, B. Z. Tang, *Angew. Chem. Int. Ed.* **2018**, *57*, 9290–9294.

[29] S.-J. Su, Y. Takahashi, T. Chiba, T. Takeda, J. Kido, *Adv. Funct. Mater.* **2009**, *19*, 1260–1267.

[30] M.-H. Chen, Y.-H. Chen, C.-T. Lin, G.-R. Lee, C.-I. Wu, D.-S. Leem, J.-J. Kim, T.-W. Pi, *J Appl. Phys.* **2009**, *105*, 113714.

[31] L. H. Smith, J. A. E. Wasey, I. D. W. Samuel, W. L. Barnes, *Adv. Funt. Mater*, **2005**, *15*, 1839–1844.

[32] T.-W. Lee, T. Noh, H.-W. Shin, O. Kwon, J.-J. Park, B.-K. Choi, M.-S. Kim, D. W. Shin, Y.-R. Kim, *Adv. Funct. Mater.* **2009**, *19*, 1625–1630.

[33] C. A. Zuniga, S. Barlow, S. R. Marder, *Chem. Mater.* **2011**, *23*, 658–681.

[34] J. W. Levell, J. P. Gunning, P. L. Burn, J. Robertson, I. D. W. Samuel, *Org. Electron.* **2010**, *11*, 1561–1568.

[35] H. Sasabe, J. Kido, *Chem. Mater.* **2011**, *23*, 621–630.

[36] M. A. Baldo, D. F. O'Brien, Y. You, A. Shoustikov, S. Sibley, M. E. Thompson, S. R. Forrest, *Nature* **1998**, *395*, 151–154.

[37] M. A. Baldo, M. E. Thompson, S. R. Forrest, *Nature* **2000**, *403*, 750–753.

[38] S. Lamansky, P. Djurovich, D. Murphy, F. Abdel-Razzaq, H.-E. Lee, C. Adachi, P. E. Burrows,§ S. R. Forrest, M. E. Thompson, *J. Am. Chem. Soc.* **2001**, *123*, 4304–4312.

[39] E. Holder, B. M. W. Langeveld, U. S. Schubert, *Adv. Mater*, **2005**, *17*, 1109–1121.

[40] Q. Wang, D. Ma, *Chem. Soc. Rev,* **2010**, *39*, 2387–2398.

[41] D. H. Kim, N. S. Cho, H. Y. Oh, J. H. Yang, W. S. Jeon, J. S. Park, M. C. Suh, J. H. Kwon, *Adv. Mater.* **2011**, *23*, 2721–2726.

[42] Y. Chi, P. T. Chou, *Chem. Soc. Rev.* **2010**, *39*, 638–655.

[43] A. K. Pal, S. Krotkus, M. Fontani, C. F. R. Mackenzie, D. B. Cordes, A. M. Z. Slawin, I. D. W. Samuel, E. Zysman-Colman, *Adv. Mater.* **2018**, *30*, 1804231.

[44] D. Volz, M. Wallesch, C. Fléchon, M. Danz, A. Verma, J. M. Navarro, D. M. Zink, S. Bräse, T. Baumann, *Green Chem.* **2015**, *17*, 1988–2011.

[45] C. Bizzarri, E. Spuling, D. M. Knoll, D. Volz, S. Bräse, *Coord. Chem. Rev.* **2018**, *373*, 49–82.

[46] C. A. Parker, C. G. Hatchard, *Trans. Faraday Soc.* **1961**, *57*, 1894–1904.

[47] P. F. Jones, A. R. Calloway, *Chem. Phys. Lett.* **1971**, *10*, 438–443.

[48]   M. Sikorski, I. V. Khmelinskii, W. Augustyniak, F. Wilkinson, *J. Chem. Soc. Faraday Trans.* **1996**, *92*, 3487–3490.

[49]   B. S. Yamanashi, D. M. Hercules, *Appl. Spectrosc.* **1971**, *25*, 457–460.

[50]   F. A. Salazar, A. Fedorov, M. N. Berberan-Santos, *Chem. Phys. Lett.* **1997**, *271*, 361–366.

[51]   H. Yersin, U. Monkowius, DE 10 2008 033 563 A1, **2008**.

[52]   A. Endo, M. Ogasawara, A. Takahashi, D. Yokoyama, Y. Kato, C. Adachi, *Adv. Mater.* **2009**, *21*, 4802–4806.

[53]   J. C. Deaton, S. C. Switalski, D. Y. Kondakov, R. H. Young, T. D. Pawlik, D. J. Giesen, S. B. Harkins, A. J. M. Miller, S. F. Mickenberg, J. C. Peters, *J. Am. Chem. Soc.* **2010**, *132*, 9499–9508.

[54]   A. Endo, K. Sato, K. Yoshimura, T. Kai, A. Kawada, H. Miyazaki, C. Adachi, *Appl. Phys. Lett.* **2011**, *98*, 083302.

[55]   H. Uoyama, K. Goushi, K. Shizu, H. Nomura, C. Adachi, *Nature* **2012**, *492*, 234–238.

[56]   Z. Yang, Z. Mao, Z. Xie, Y. Zhang, S. Liu, J. Zhao, J. Xu, Z. Chi, M. P. Aldred, *Chem. Soc. Rev.* **2017**, *46*, 915–1016.

[57]   M. Y. Wong, E. Zysman-Colman, *Adv. Mater.* **2017**, *29*. 1605444.

[58]   Y. Im, M. Kim, Y. J. Cho, J.-A. Seo, K. S. Yook, J. Y. Lee, *Chem. Mater.* **2017**, *29*, 1946–1963.

[59]   Y. Liu, C. Li, Z. Ren, S. Yan, M. R. Bryce, *Nat. Rev. Mater.* **2018**, *3*, 18020.

[60]   J. Lee, K. Shizu, H. Tanaka, H. Nakanotani, T. Yasuda, H. Kaji, C. Adachi, *J Mater. Chem. C* **2015**, *3*, 2175–2181.

[61]   I. S. Park, S. Y. Lee, C. Adachi, T. Yasuda, *Adv. Funct. Mater.* **2016**, *26*, 1813–1821.

[62]   X. Cai, X. Li, G. Xie, Z. He, K. Gao, K. Liu, D. Chen, Y. Cao, S. J. Su, *Chem. Sci* **2016**, *7*, 4264–4275.

[63]   S. Y. Lee, T. Yasuda, H. Komiyama, J. Lee, C. Adachi, *Adv. Mater.* **2016**, *28*, 4019–4024.

[64]   K. Shizu, Y. Sakai, H. Tanaka, S. Hirata, C. Adachi, H. Kaji, *ITE Trans. Media Technol. Appl.* **2015**, *3*, 108.

[65]   M. Moral, L. Muccioli, W. J. Son, Y. Olivier, J. C. Sancho-Garcia, *J. Chem. Theory Comput.* **2015**, *11*, 168–177.

[66]   K. Albrecht, K. Matsuoka, K. Fujita, K. Yamamoto, *Angew. Chem. Int. Ed.* **2015**, *54*, 5677–5682.

[67]   P. Data, P. Pander, M. Okazaki, Y. Takeda, S. Minakata, A. P. Monkman, *Angew. Chem. Int. Ed.* **2016**, *55*, 5739–5744

[68]   T. Takahashi, K. Shizu, T. Yasuda, K. Togashi, C. Adachi, *Sci. Technol. Adv. Mater.* **2014**, *15*, 034202.

[69]   Y. Zhang, D. Zhang, M. Cai, Y. Li, D. Zhang, Y. Qiu, L. Duan, *Nanotechnology* **2016**, *27*, 094001.

[70]   W. L. Tsai, M. H. Huang, W. K. Lee, Y. J. Hsu, K. C. Pan, Y. H. Huang, H. C. Ting, M. Sarma, Y. Y. Ho, H. C. Hu, C. C. Chen, M. T. Lee, K. T. Wong, C. C. Wu, *Chem. Commun.* **2015**, *51*, 13662–13665.

[71]   M. Kim, S. K. Jeon, S. H. Hwang, J. Y. Lee, *Adv. Mater.* **2015**, *27*, 2515–2520.

[72]   S. Y. Lee, T. Yasuda, I. S. Park, C. Adachi, *Dalton. Trans.* **2015**, *44*, 8356–8359.

[73]   I. H. Lee, W. Song, J. Y. Lee, *Org. Electron.* **2016**, *29*, 22–26.

[74]   K. Suzuki, S. Kubo, K. Shizu, T. Fukushima, A. Wakamiya, Y. Murata, C. Adachi, H. Kaji, *Angew. Chem. Int. Ed.* **2015**, *54*, 15231–15235.

[75]   Y. J. Cho, S. K. Jeon, J. Y. Lee, *Adv. Optical Mater.* **2016**, *4*, 688–693.

[76]   L. S. Cui, H. Nomura, Y. Geng, J. U. Kim, H. Nakanotani, C. Adachi, *Angew. Chem. Int. Ed.* **2017**, *56*, 1571–1575.

[77]   W. Huang, M. Einzinger, T. Zhu, H. S. Chae, S. Jeon, S.-G. Ihn, M. Sim, S. Kim, M. Su, G. Teverovskiy, T. Wu, T. Van Voorhis, T. M. Swager, M. A. Baldo, S. L. Buchwald, *Chem. Mater.* **2018**, *30*, 1462–1466.

[78]   N. Sharma, E. Spuling, C. M. Mattern, W. Li, O. Fuhr, Y. Tsuchiya, C. Adachi, S. Bräse, I. D. W. Samuel, E. Zysman-Colman, *Chem. Sci.* **2019**, *10*, 6689–6696.

[79]   P. Rajamalli, N. Senthilkumar, P. Gandeepan, P. Y. Huang, M. J. Huang, C. Z. Ren-Wu, C. Y. Yang, M. J. Chiu, L. K. Chu, H. W. Lin, C. H. Cheng, *J. Am. Chem. Soc.* **2016**, *138*, 628–634.

[80]   H. Tsujimoto, D. G. Ha, G. Markopoulos, H. S. Chae, M. A. Baldo, T. M. Swager, *J. Am. Chem. Soc.* **2017**, *139*, 4894–4900.

[81]   D. J. Cram, J. M. Cram, *Acc. Chem. Res.* **1971**, *4*, 204–213.

[82]   J. Zyss, I. Ledoux, S. Volkov, V. Chernyak, S. Mukamel, G. P. Bartholomew, G. C. Bazan, *J. Am. Chem. Soc.* **2000**, *122*, 11956–11962.

[83]   G. P. Bartholomew, G. C. Bazan, *Acc. Chem. Res.* **2001**, *34*, 30–39.

[84]   J. W. Hong, H. Y. Woo, Bin Liu, G. C. Bazan, *J. Am. Chem. Soc.* **2005**, *127*, 7435–7443.

[85]   Y. Morisaki, Y. Chujo, *Polym. Chem.* **2011**, *2*, 1249–1257.

[86]   Y. Morisaki, Y. Chujo, *Chem. Lett.* **2012**, *41*, 840–846.

[87]   E. Spuling, N. Sharma, I. D. W. Samuel, E. Zysman-Colman, S. Bräse, *Chem. Commun.* **2018**, *54*, 9278–9281.

[88]   M. Y. Zhang, Z. Y. Li, B. Lu, Y. Wang, Y. D. Ma, C. H. Zhao, *Org. Lett.* **2018**, *20*, 6868–6871.

[89]   Y. Z. Shi, K. Wang, X. Li, G. L. Dai, W. Liu, K. Ke, M. Zhang, S. L. Tao, C. J. Zheng, X. M. Ou, X. H. Zhang, *Angew. Chem. Int. Ed.* **2018**, *57*, 9480–9484.

[90]   T. Hatakeyama, K. Shiren, K. Nakajima, S. Nomura, S. Nakatsuka, K. Kinoshita, J. Ni, Y. Ono, T. Ikuta, *Adv. Mater.* **2016**, *28*, 2777–2781.

[91]   Y. Yuan, X. Tang, X. Y. Du, Y. Hu, Y. J. Yu, Z. Q. Jiang, L. S. Liao, S. T. Lee, *Adv. Optical Mater.* **2019**, *7*, 1801536.

[92]   D. Hall, S. M. Suresh, P. L. dos Santos, E. Duda, S. Bagnich, A. Pershin, P. Rajamalli, D. B. Cordes, A. M. Z. Slawin, D. Beljonne, A. Köhler, I. D. W. Samuel, Y. Olivier, E. Zysman-Colman, *Adv. Optical Mater.* **2019**, *8*, 1901627.

[93]   D. R. Lee, M. Kim, S. K. Jeon, S. H. Hwang, C. W. Lee, J. Y. Lee, *Adv. Mater.* **2015**, *27*, 5861–5867.

[94]   Y. Liu, G. Xie, K. Wu, Z. Luo, T. Zhou, X. Zeng, J. Yu, S. Gong, C. Yang, *J. Mater. Chem. C* **2016**, *4*, 4402–4407.

[95]   D. Zhang, X. Cao, Q. Wu, M. Zhang, N. Sun, X. Zhang, Y. Tao, *J. Mater. Chem. C* **2018**, *6*, 3675–3682.

[96]   Y. Sun, S. R. Forrest, *Nat. Photonics* **2008**, *2*, 483–487.

[97]   K. Saxena, D. S. Mehta, V. K. Rai, R. Srivastava, G. Chauhan, M. N. Kamalasanan, *J. Lumin.* **2008**, *128*, 525–530.

[98]   K. Hong, H. K. Yu, I. Lee, K. Kim, S. Kim, J. L. Lee, *Adv. Mater.* **2010**, *22*, 4890–4894.

[99]   C. Y. Chen, W. K. Lee, Y. J. Chen, C. Y. Lu, H. Y. Lin, C. C. Wu, *Adv. Mater.* **2015**, *27*, 4883–4888.

[100]   L. Kinner, S. Nau, K. Popovic, S. Sax, I. Burgués-Ceballos, F. Hermerschmidt, A. Lange, C. Boeffel, S. A. Choulis, E. J. W. List-Kratochvil, *Appl. Phys. Lett.* **2017**, *110*, 101107.

[101]   X. Liang, Z. L. Tu, Y. X. Zheng, *Chem. Eur. J.* **2019**, *25*, 5623–5642.

[102]   Q. Zhang, B. Li, S. Huang, H. Nomura, H. Tanaka, C. Adachi, *Nat. Photonics* **2014**, *8*, 326–332.

[103]   I. S. Park, K. Matsuo, N. Aizawa, T. Yasuda, *Adv. Funct. Mater.* **2018**, *28*, 1802031.

[104]   K. H. Kim, J. J. Kim, *Adv. Mater.* **2018**, *30*, 1705600.

[105]   Y. Watanabe, H. Sasabe, J. Kido, *Bull. Chem. Soc. Jpn.* **2019**, *92*, 716–728.

[106]   T. Marcato, C. J. Shih, *Helv. Chim. Acta.* **2019**, *102*, 1900048.

[107]   M. Liu, R. Komatsu, X. Cai, K. Hotta, S. Sato, K. Liu, D. Chen, Y. Kato, H. Sasabe, S. Ohisa, Y. Suzuri, D. Yokoyama, S.-J. Su, J. Kido, *Chem. Mater.* **2017**, *29*, 8630–8636.

[108]   H. Kaji, H. Suzuki, T. Fukushima, K. Shizu, K. Suzuki, S. Kubo, T. Komino, H. Oiwa, F. Suzuki, A. Wakamiya, Y. Murata, C. Adachi, *Nat. Commun.* **2015**, *6*, 8476.

[109]   T. A. Lin, T. Chatterjee, W. L. Tsai, W. K. Lee, M. J. Wu, M. Jiao, K. C. Pan, C. L. Yi, C. L. Chung, K. T. Wong, C. C. Wu, *Adv. Mater.* **2016**, *28*, 6976–6983.

[110]   S. Xiang, X. Lv, S. Sun, Q. Zhang, Z. Huang, R. Guo, H. Gu, S. Liu, L. Wang, *J. Mater. Chem. C* **2018**, *6*, 5812–5820.

[111]   W. Li, B. Li, X. Cai, L. Gan, Z. Xu, W. Li, K. Liu, D. Chen, S. J. Su, *Angew. Chem. Int. Ed.* **2019**, *58*, 11301–11305.

[112]   L. Ji, S. Griesbeck, T. B. Marder, *Chem. Sci.* **2017**, *8*, 846–863.

[113]   Y. Kitamoto, T. Namikawa, D. Ikemizu, Y. Miyata, T. Suzuki, H. Kita, T. Sato, S. Oi, *J. Mater. Chem. C* **2015**, *3*, 9122–9130.

[114]   T. Katayama, S. Nakatsuka, H. Hirai, N. Yasuda, J. Kumar, T. Kawai, T. Hatakeyama, *J. Am. Chem. Soc.* **2016**, *138*, 5210–5213.

[115]   X. Y. Wang, A. Narita, W. Zhang, X. Feng, K. Müllen, *J. Am. Chem. Soc.* **2016**, *138*, 9021–9024.

[116]   K. Matsuo, T. Yasuda, *Chem. Commun.* **2019**, *55*, 2501–2504.

[117]   H. Shi, J. Yuan, X. Wu, X. Dong, L. Fang, Y. Miao, H. Wang, F. Cheng, *New J. Chem.* **2014**, *38*, 2368–2378.

[118]   K. Shizu, H. Noda, H. Tanaka, M. Taneda, M. Uejima, T. Sato, K. Tanaka, H. Kaji, C. Adachi, *J. Phys. Chem. C* **2015**, *119*, 26283–26289.

[119]   K. Shizu, H. Tanaka, M. Uejima, T. Sato, K. Tanaka, H. Kaji, C. Adachi, *J. Phys. Chem. C* **2015**, *119*, 1291–1297.

[120]   G. Xie, X. Li, D. Chen, Z. Wang, X. Cai, D. Chen, Y. Li, K. Liu, Y. Cao, S. J. Su, *Adv. Mater.* **2016**, *28*, 181–187.

[121]   Z. Zhang, J. Xie, Z. Wang, B. Shen, H. Wang, M. Li, J. Zhang, J. Cao, *J. Mater. Chem. C* **2016**, *4*, 4226–4235.

[122]   Z. Zhang, J. Xie, H. Wang, B. Shen, J. Zhang, J. Hao, J. Cao, Z. Wang, *Dyes Pigm.* **2016**, *125*, 299–308.

[123]   M. Numano, N. Nagami, S. Nakatsuka, T. Katayama, K. Nakajima, S. Tatsumi, N. Yasuda, T. Hatakeyama, *Chem. Eur. J.* **2016**, *22*, 11574–11577.

[124]   Y. H. Lee, S. Park, J. Oh, J. W. Shin, J. Jung, S. Yoo, M. H. Lee, *ACS Appl. Mater. Interfaces* **2017**, *9*, 24035–24042.

[125]   Y. H. Lee, S. Park, J. Oh, S.-J. Woo, A. Kumar, J.-J. Kim, J. Jung, S. Yoo, M. H. Lee, *Adv. Optical Mater.* **2018**, *6*, 1800385.

[126]   X. L. Chen, J. H. Jia, R. Yu, J. Z. Liao, M. X. Yang, C. Z. Lu, *Angew. Chem. Int. Ed.* **2017**, *56*, 15006–15009.

[127]   A. Kumar, J. Oh, J. Kim, J. Jung, M. H. Lee, *Dyes Pigm.* **2019**, *168*, 273–280.

[128]   H.-p. Shi, J.-x. Dai, X.-h. Wu, L.-w. Shi, J.-d. Yuan, L. Fang, Y.-q. Miao, X.-g. Du, H. Wang, C. Dong, *Org. Electron.* **2013**, *14*, 868–874.

[129]   C. Sanchez, C. Mendez, J. A. Salas, *Nat. Prod. Rep.* **2006**, *23*, 1007–1045.

[130]   T. Janosik, A. Rannug, U. Rannug, N. Wahlstrom, J. Slatt, J. Bergman, *Chem. Rev.* **2018**, *118*, 9058–9128.

[131]   K. Sato, K. Shizu, K. Yoshimura, A. Kawada, H. Miyazaki, C. Adachi, *Phys. Rev. Lett.* **2013**, *110*, 247401.

[132]   D. Zhang, X. Song, M. Cai, H. Kaji, L. Duan, *Adv. Mater.* **2018**, *30, 1705406*.

[133]   Y. J. Kang, J. Y. Lee, *Dyes Pigm.* **2017**, *138*, 176–181.

[134]   Y. Li, J.-J. Liang, H.-C. Li, L.-S. Cui, M.-K. Fung, S. Barlow, S. R. Marder, C. Adachi, Z.-Q. Jiang, L.-S. Liao, *J. Mater. Chem. C* **2018**, *6*, 5536–5541.

[135]   U. Pindur, J. Mueller, *Arch. Pharm.* **1987**, *320*, 280–282.

[136]   H. K. Panesar, J. Solano, T. G. Minehan, *Org. Biomol. Chem.* **2015**, *13*, 2879–2883.

[137]   Q. Zhang, H. Zhuang, J. He, S. Xia, H. Li, N. Li, Q. Xu, J. Lu, *J. Mater. Chem. C* **2015**, *3*, 6778–6785.

[138]   S. Ameen, J. Lee, H. Han, M. C. Suh, C. Lee, *RSC Adv.* **2016**, *6*, 33212–33220.

[139]   Y. Gao, A. Hlil, J. Wang, K. Chen, A. S. Hay, Macromolecules **2007**, *40*, 4744–4746.

[140]   B. Li, H. Nomura, H. Miyazaki, Q. Zhang, K. Yoshida, Y. Suzuma, A. Orita, J. Otera, C. Adachi, *Chem. Lett.* **2014**, *43*, 319–321.

[141]   M. Kim, S. K. Jeon, S.-H. Hwang, S.-S. Lee, E. Yu, J. Y. Lee, *J. Phys. Chem. C* **2016**, *120*, 2485–2493.

[142]   M. Usui, D. T. Fukushima, S. Tanaka, WO2016002646, **2016**.

[143]   L. Li, J. Xiang, C. Xu, *Org. Lett.* **2007**, *9*, 4877–4879.

[144]   Z. Hassan, E. Spuling, D. M. Knoll, J. Lahann, S. Brase, *Chem. Soc. Rev.* **2018**, *47*, 6947–6963.

[145]   Z. Hassan, E. Spuling, D. M. Knoll, S. Bräse, *Angew. Chem. Int. Ed.* **2020**, *59*, 2156–2170.

[146]   F. Vögtle, P. Neumann, *Tetrahedron Lett.* **1969**, *60*, 5329–5334.

[147]   R. S. Cahn, C. Ingold, V. Prelog, *Angew. Chem. Int. Ed.* **1966**, *5*, 385–415.

[148]   Y. Hu, *Metal Catalyzed N–H and O–H Insertion from α-Diazocarbonyl Compounds, Logos-Verlag.* **2019**.

[149]   M. Yokoyama, K. Inada, Y. Tsuchiya, H. Nakanotani, C. Adachi, *Chem. Commun.* **2018**, *54*, 8261–8264.

[150]   C. Y. Chan, L. S. Cui, J. U. Kim, H. Nakanotani, C. Adachi, *Adv. Funct. Mater.* **2018**, *28*, 1706023.

[151]   P. Lennartz, G. Raabe, C. Bolm, *Isr. J. Chem.* **2012**, *52*, 171–179.

[152]   S. L. Buchwald, W. Huang, WO2016196885 A1, **2016**.

[153]   J. T. Je, S. H. Kim, A. R. Choi, KR 2010055351, **2010**.

[154]   K. J. Yoon, H. J. Noh, D. W. Yoon, I. A. Shin, J. Y. Kim, EP3029753 A1, **2016**.

[155]   H.-F. Chen, L.-C. Chi, W.-Y. Hung, W.-J. Chen, T.-Y. Hwu, Y.-H. Chen, S.-H. Chou, E. Mondal, Y.-H. Liu, K.-T. Wong, *Org. Electron.* **2012**, *13*, 2671–2681.

[156]   S. Wu, Y. Liu, G. Yu, J. Guan, C. Pan, Y. Du, X. Xiong, Z. Wang, *Macromolecules* **2014**, *47*, 2875–2882.

[157]   W. Yu, S. Gu, Y. Fu, S. Xiong, C. Pan, Y. Liu, G. Yu, *J. Catal.* **2018**, *362*, 1–9.

[158]   Y. Im, S. H. Han, J. Y. Lee, *J. Mater. Chem. C* **2018**, *6*, 5012–5017.

[159]   K. A. McReynolds, R. S. Lewis, L. K. G. Ackerman, G. G. Dubinina, W. W. Brennessel, D. A. Vicic, *J. Fluorine Chem.* **2010**, *131*, 1108–1112.

[160]   T. Schareina, X.-F. Wu, A. Zapf, A. Cotté, M. Gotta, M. Beller, *Top. Catal.* **2012**, *55*, 426–431.

[161]   R. Nagarajan, D. Sreenivas, *Synlett.* **2012**, *23*, 1007–1012.

[162]   Y. Liu, X. Ye, G. Liu, Y. Lv, X. Zhang, S. Chen, J. W. Y. Lam, H. S. Kwok, X. Tao, B. Z. Tang, *J. Mater. Chem. C* **2014**, *2*, 1004–1009.

[163]   H. Shi, M. Li, D. Xin, L. Fang, J. Roose, H. Peng, S. Chen, B. Z. Tang, *Dyes Pigm.* **2016**, *128*, 304–313.

[164]   S. S. Chirke, J. S. Krishna, B. B. Rathod, S. R. Bonam, V. M. Khedkar, B. V. Rao, H. M. Sampath Kumar, P. R. Shetty, *ChemistrySelect* **2017**, *2*, 7309–7318.

[165]   H. Shi, X. Zhang, C. Gui, S. Wang, L. Fang, Z. Zhao, S. Chen, B. Z. Tang, *J. Mater. Chem. C* **2017**, *5*, 11741–11750.

[166]   K.-C. Pan, S.-W. Li, Y.-Y. Ho, Y.-J. Shiu, W.-L. Tsai, M. Jiao, W.-K. Lee, C.-C. Wu, C.-L. Chung, T. Chatterjee, Y.-S. Li, K.-T. Wong, H.-C. Hu, C.-C. Chen, M.-T. Lee, *Adv. Funct. Mater.* **2016**, *26*, 7560–7571.

[167]   R. Gomez-Bombarelli, J. Aguilera-Iparraguirre, T. D. Hirzel, D. Duvenaud, D. Maclaurin, M. A. Blood-Forsythe, H. S. Chae, M. Einzinger, D. G. Ha, T. Wu, G. Markopoulos, S. Jeon, H. Kang, H. Miyazaki, M. Numata, S. Kim, W. Huang, S. I. Hong, M. Baldo, R. P. Adams, A. Aspuru-Guzik, *Nat. Mater.* **2016**, *15*, 1120–1127.

[168]   Y. Wada, S. Kubo, H. Kaji, *Adv. Mater.* **2018**, *30*, *1705641*.

[169]   C. Li, C. Duan, C. Han, H. Xu, *Adv. Mater.* **2018**, *30*, 1804228.

[170]   X. Liang, Z. P. Yan, H. B. Han, Z. G. Wu, Y. X. Zheng, H. Meng, J. L. Zuo, W. Huang, *Angew. Chem. Int. Ed.* **2018**, *57*, 11316–11320.

[171]   T.-L. Wu, M.-J. Huang, C.-C. Lin, P.-Y. Huang, T.-Y. Chou, R.-W. Chen-Cheng, H.-W. Lin, R.-S. Liu, C.-H. Cheng, *Nat. Photonics* **2018**, *12*, 235–240.

[172]   S. Hirata, Y. Sakai, K. Masui, H. Tanaka, S. Y. Lee, H. Nomura, N. Nakamura, M. Yasumatsu, H. Nakanotani, Q. Zhang, K. Shizu, H. Miyazaki, C. Adachi, *Nat. Mater.* **2015**, *14*, 330–336.

[173]   M.-L. Louillat, F. W. Patureau, *Org. Lett,* **2013**, *15*, 164–167.

[174]   A. Tomkeviciene, J. V. Grazulevicius, D. Volyniuk, V. Jankauskas, G. Sini, *Phys. Chem. Chem. Phys.* **2014**, *16*, 13932–13942.

[175]   D. T. Yang, T. Nakamura, Z. He, X. Wang, A. Wakamiya, T. Peng, S. Wang, *Org. Lett.* **2018**, *20*, 6741–6745.

[176]   P. Rajamalli, N. Senthilkumar, P. Y. Huang, C. C. Ren-Wu, H. W. Lin, C. H. Cheng, *J. Am. Chem. Soc.* **2017**, *139*, 10948–10951.

[177]   P. Ganesan, D.-G. Chen, J.-L. Liao, W.-C. Li, Y.-N. Lai, D. Luo, C.-H. Chang, C.-L. Ko, W.-Y. Hung, S.-W. Liu, G.-H. Lee, P.-T. Chou, Y. Chi, *J. Mater. Chem. C* **2018**, *6*, 10088–10100.

[178]   Q. Zhang, S. Sun, X. Lv, W. Liu, H. Zeng, R. Guo, S. Ye, P. Leng, S. Xiang, L. Wang, *Mater. Chem. Front.* **2018**, *2*, 2054–2062.

[179]   P. L. Dos Santos, D. Chen, P. Rajamalli, T. Matulaitis, D. B. Cordes, A. M. Z. Slawin, D. Jacquemin, E. Zysman-Colman, I. D. W. Samuel, *ACS Appl. Mater. Interfaces* **2019**, *11*, 45171–45179.

[180]   L. Wang, X. Cai, B. Li, M. Li, Z. Wang, L. Gan, Z. Qiao, W. Xie, Q. Liang, N. Zheng, K. Liu, S. J. Su, *ACS Appl. Mater. Interfaces* **2019**, *11*, 45999–46007.

[181]   T. Imagawa, S. Hirata, K. Totani, T. Watanabe, M. Vacha, *Chem. Commun.* **2015**, *51*, 13268–13271.

[182]   S. Feuillastre, M. Pauton, L. Gao, A. Desmarchelier, A. J. Riives, D. Prim, D. Tondelier, B. Geffroy, G. Muller, G. Clavier, G. Pieters, *J. Am. Chem. Soc.* **2016**, *138*, 3990–3993.

[183]   M. Li, S. H. Li, D. Zhang, M. Cai, L. Duan, M. K. Fung, C. F. Chen, *Angew. Chem. Int. Ed.* **2018**, *57*, 2889–2893.

[184]   D. W. Zhang, M. Li, C. F. Chen, *Chem. Soc. Rev.* **2020**, DOI: 10.1039/c9cs00680j.

[185]   X. Liang, T. T. Liu, Z. P. Yan, Y. Zhou, J. Su, X. F. Luo, Z. G. Wu, Y. Wang, Y. X. Zheng, J. L. Zuo, *Angew. Chem. Int. Ed.* **2019**, *58*, 17220–17225.

[186]   Z. Wang, Y. Li, X. Cai, D. Chen, G. Xie, K. Liu, Y. C. Wu, C. C. Lo, A. Lien, Y. Cao, S. J. Su, *ACS Appl. Mater. Interfaces* **2016**, *8*, 8627–8636.

[187]   C. S. Wang, Y. C. Wei, K. H. Chang, P. T. Chou, Y. T. Wu, *Angew. Chem. Int. Ed.* **2019**, *58*, 10158–10162.

8 Appendix

8.1 Curriculum Vitae

Personal Details

Name	Zhen Zhang
Address	Kaiserstr. 29
	76131 Karlsruhe
	Germany
Telephone	+49 17630115512
E-mail	zhenzhang2015@outlook.com
Date of Birth	30.05.1988
Nationality	Chinese

Education

10.2016 – 04.2020	PhD Student in Organic Chemistry,
	Karlsruhe Institute of Technology, Karsruhe, Germany
	Thesis Advisor: Prof. Dr. Stefan Bräse
	Thesis: *Design and Synthesis of N-Heterocyclic Donor Based TADF Emitters for Application in OLEDs*
09.2013 – 06.2016	Master of Organic Chemistry
	Shanghai University, Shanghai, China
	Thesis Advisor: Assoc. Prof. Dr. Zixing Wang
	Thesis: *Synthesis and Optoelectronic Properties of Nitrogen Heterocycle Host Materials and Red Iridium Complexes*
09.2009 – 06.2013	Bachelor of Applied Chemistry
	Jiangxi Science and Technology Normal University, Nanchang, China
	Thesis Advisor: Prof. Dr. Qiang Xiao
	Thesis: *Synthesis of the Key intermediate of a Marine Alkaloid-Rigidin Fluorinated Analogues*

Scholarships/Awards

05.2016	Scholarship from China Scholarship Council

8.2 Publications and Conference Contributions

Publications

1. <u>Zhen Zhang</u>, Shiv Kumar, Sergey Bagnich, Eduard Spuling, Fabian Hundemer, Martin Nieger, Zahid Hassan, Anna Köhler, Eli Zysman-Colman, Stefan Bräse, *Submitted to Front. Chem.* **2020**.

OBO-Fused Benzo[fg]tetracene as Acceptor with Potential for Thermally Activated Delayed Fluorescence Emitters

2. <u>Zhen Zhang</u>, Ettore Crovini, Paloma L. dos Santos, Bilal A. Naqvi, David B. Cordes, Alexandra M. Z. Slawin, Prakhar Sahay, Wolfgang Brütting, Ifor D. W. Samuel, Stefan Bräse, Eli Zysman-Colman, *Submitted to Adv. Optical Mater.* **2020**.

Efficient Sky-Blue Organic Light-Emitting Diodes Using a Highly Horizontally Oriented Thermally Activated Delayed Fluorescence Emitter

3. <u>Zhen Zhang</u>, Zahid Hassan, Olaf Fuhr, Stefan Bräse, *Submitted to Chem. Eur. J.* **2020**.

Molecular Design and Synthesis of Novel Indolo[3,2-b]carbazoles with Methyl-Substituted π-Bridges: Tuning the Behavior of TADF through Changing Electron-Withdrawing Strength of Acceptors

4. Bilal A. Naqvi, Markus Schmid, Ettore Crovini, Prakhar Sahay, Tassilo Naujoks, Francesco Rodella, <u>Zhen Zhang</u>, Peter Strohriegl, Stefan Bräse, Eli Zysman-Colman, Wolfgang Bruetting, *Front. Chem.* **2020**, *DOI: 10.3389/fchem.2020.00750.*

What Controls the Orientation of TADF Emitters?

5. <u>Zhen Zhang</u>, Abhishek Kumar Gupta, Eduard Spuling, Eli Zysman-Colman, Stefan Bräse, *In preparation.*

Electron-Withdrawing Groups Modified Carbazolophane Donors for Deep Blue Thermally Activated Delayed Fluorescence OLEDs

6. <u>Zhen Zhang</u>, Stefan Diesing, Ettore Crovini, Zahid Hassan, Ifor D. W. Samuel, Eli Zysman-Colman, Stefan Bräse, *In preparation.*

Novel Centrosymmetric Through-Space Donors for Solution-Processed Blue Thermally Activated Delayed Fluorescence OLEDs

7. **Zhen Zhang**, Eimantas Duda, Ettore Crovini, Anna Köhler, Eli Zysman-Colman, Stefan Bräse, *In preparation.*

Highly Efficient Solution-Processed OLEDs Enhanced by TADF Emitters with Short Delayed Fluorescence Lifetime

8. Peng Wu, Jun Zhu, **Zhen Zhang**, Dehai Dou, Hedan Wang, Bin Wei, Zixing Wang, *Dyes Pigm.* **2018**, *156*, 185–191.

Nitrogen Introduction of Spirobifluorene to Form α-, β-, γ-, and δ-Aza-9,9'- spirobifluorenes: New Bipolar System for Efficient Blue Organic Light-Emitting Diodes

9. Zixing Wang, Jun Zhu, Zhiwei Liu, Peng Wu, Hedan Wang, **Zhen Zhang**, Bin Wei, *J. Mater. Chem. C* **2017**, 5, 6982–6988.

Thermally Activated Delayed Fluorescence of Co-Deposited Copper(I) Complexes: Cost-Effective Emitters for Highly Efficient Organic Light-Emitting Diodes

10. **Zhen Zhang**, Jingwei Xie, Zixing Wang, Bowen Shen, Hedan Wang, Minjie Li, Jianhua Zhang, Jin Cao, *J. Mater. Chem. C* **2016**, *4*, 4226–4235.

Manipulation of Electron Deficiency of δ-Carboline Derivatives as Bipolar Hosts for Blue Phosphorescent Organic Light-Emitting Diodes with High Efficiency at 1000 cd m^{-2}

11. **Zhen Zhang**, Jingwei Xie, Huibin Wang, Bowen Shen, Jianhua Zhang, Jian Hao, Jin Cao, Zixing Wang, *Dyes Pigm.* **2016**,*125*, 299–308.

Synthesis, Photophysical and Optoelectronic Properties of Quinazolinecentered Dyes and their Applications in Organic Light-Emitting Diodes

Posters

1. **Zhen Zhang**, Stefan Seifermann, Daniel Volz, Stefan Bräse. 7th of September 2017, Frankfurt am Main, 1st International TADF Symposium.

Three-Coordinate Boron TADF Compounds for OLEDs

8.3 Acknowledgements

First of all, I would like to express my deepest gratitude to my advisor Prof. Dr. Stefan Bräse for his guidance and patience during my whole Ph.D. study. I want to thank him for his continuous support and valuable suggestions on my research work. Also, I appreciate the encouragements and opportunities that I was given by him for my academic and personal development.

I am very grateful to Prof. Dr. Eli Zysman-Colman at the University of St. Andrews for the fruitful collaboration to deeply understand the photophysical aspects. I appreciate the help from him and his students, Ettore Crovini, Stefan Diesing, Dr. Shiv Kumar, Dr. Eduard Spuling and Dr. Abhishek Kumar Gupta for the theoretical calculations and photophysical investigations of the luminescent emitters I have provided.

Furthermore, I want to thank Prof. Dr. Ifor D. W. Samuel and Dr. Paloma L. dos Santos at the University of St. Andrews for the device fabrication of the indolocarbazole project.

I would like to thank Prof. Dr. Anna Köhler and Eimantas Duda at the University of Bayreuth for the photophysical measurements of the triarylboron-based emitters. Many thanks also to Prof. Dr. Wolfgang Bruetting and Bilal Abbas Naqvi at the University of Augsburg for the orientation studies of the indolocarbazole emitters.

I would like to thank Dr. Stefan Seifermann and Nico-Patrick Thöbes from cynora GmbH who helped me design the molecules and do the measurements of OBO-based emitters.

Furthermore, I want to acknowledge the work of Dr. Olaf Fuhr for crystal structure analyses at KIT, Dr. Martin Nieger for crystal structure analyses at the University of Helsinki, Dr. Andreas Rapp and Tanja Ohmer-Scherrer for NMR spectroscopy, Dr. Norbert Foitzik, Lara Hirsch and Angelika Mösle for mass spectrometry and IR, Dr. Karolin Kohnle for elemental analysis. All of them helped me very kindly by analyzing all compounds and without whose work this thesis would not have been possible.

I kindly acknowledge Prof. Dr. Uli Lemmer for the acceptance of the co-reference of this thesis.

I also want to especially thank Dr. Eduard Spuling, Jasmin Busch and Gloria Hong for taking the time to proof read and give me very helpful remarks and suggestions.

I would like to thank the whole AK Bräse group for this good time. It is my pleasure to be part of the team. In particular, many thanks go to Dr. Yuling Hu, Dr. Anne Schneider, Dr. Stephan Münch, Dr. Robin Bär, Dr Johannes Karcher, Susanne Kirchner, Sarah Al Muthafer and Nicolai Rosenbaum. A huge thanks to Dr. Christin Bednarek, Christiane Lampert, Janine Bolz and Selin Samur, whose help cannot be overestimated in keeping the group running.

I also want to thank the guys in TADF and Paracyclophane subgroup, Dr. Zahid Hassan, Dr. Fabian Hundemer, Jasmin Busch, Xuemin Gan, Maria Kaczmarek, Gloria Hong, Céline Leonhardt, Dr. Daniel Knoll, Christoph Zippel and Yichuan Wang for their helps on my research work.

I would like to thank for the company from my dear Chinese colleagues and friends, Dr. Yuling Hu, Xuemin Gan, Yichuan Wang, Dr. Haixia Wang, Dr. Xiaoqian Ge, Ningyi Li, who made my time in Karlsruhe very enjoyable.

A particular and very big thank you goes to Dr. Eduard Spuling whose guidance along my study in the Bräse group needs to be emphasized deeply. He has always been a great ideas maker for many issues and a mentor with humorous but always deliberate suggestions.

The biggest thanks belong to my parents, my family and my fiancee Pingping Tian. Your support and understanding give me the motivation and courage to study hard and go further. Your expectations and encouragements give me the determination and strength to continuously improve.